```
620.00425 K15i              7085722
KATZ, RANDY H., 1955-
INFORMATION MANAGEMENT FOR
ENGINEERING DESIGN
```

**DO NOT REMOVE
CARDS FROM POCKET**

ALLEN COUNTY PUBLIC LIBRARY

FORT WAYNE, INDIANA 46802

You may return this book to any agency, branch,
or bookmobile of the Allen County Public Library.

DEMCO

Surveys in Computer Science

Editors:

G. Schlageter
F. Stetter
E. Goto
J. Hartmanis

Randy H. Katz

Information Management for Engineering Design

With 33 Figures

Springer-Verlag
Berlin Heidelberg New York Tokyo

Prof. R.H. Katz
Computer Science Division
Electrical Engineering and
Computer Science Department
University of California, Berkeley
Berkeley, CA 94720
USA

Allen County Public Library
Ft. Wayne, Indiana

ISBN 3-540-15130-3
Springer-Verlag Berlin Heidelberg New York Tokyo
ISBN 0-387-15130-3
Springer-Verlag New York Heidelberg Berlin Tokyo

Library of Congress Cataloging in Publication Data. Katz, R.H. (Randy H.), 1955. Information management for engineering design (Surveys in computer science). Bibliography: p. 1. Engineering design-Data processing. 2. Computer-aided design. I. Title. II. Series. TA174.K38 1985 620'.00425'0285 85-9932. ISBN 0-387-15130-3 (U.S.)

This work is subject to copyright. All rights are reserved, whether the whole or part of the material is concerned, specifically those of translation, reprinting, re-use of illustrations, broadcasting, reproduction by photocopying machine or similar means, and storage in data banks. Under § 54 of the German Copyright Law where copie are made for other than private use a fee is payable to "Verwertungsgesellschaft Wort", Munich.

© Springer-Verlag Berlin Heidelberg 1985
Printed in Germany

The use of registered names, trademarks, etc. in this publication does not imply, even in the absence of a specific statement, that such names are exempt from the relevant protective laws and regulations and therefore free for general use.

Printing: betz-druck, Darmstadt; Bookbinding: J. Schäffer OHG, Grünstadt
2145/3140-543210

Preface

Computer-aided design systems have become a big business. Advances in technology have made it commercially feasible to place a powerful engineering workstation on every designer's desk. A major selling point for these workstations is the computer-aided design software they provide, rather than the actual hardware. The trade magazines are full of advertisements promising full menu design systems, complete with an integrated database (preferably "relational"). What does it all mean?

This book focuses on the critical issues of managing the information about a large design project. While undeniably one of the most important areas of CAD, it is also one of the least understood. Merely glueing a database system to a set of existing tools is not a solution. Several additional system components must be built to create a true design management system. These are described in this book.

The book has been written from the viewpoint of how and when to apply database technology to the problems encountered by builders of computer-aided design systems. Design systems provide an excellent environment for discovering how far we can generalize the existing database concepts for non-commercial applications. This has emerged as a major new challenge for database system research. We have attempted to avoid a "database egocentric" view by pointing out where existing database technology is inappropriate for design systems, at least given the current state of the database art.

Acknowledgements. The intellectual father of the work reported herein is Paul Losleben, who introduced me to the problems of databases for CAD through his papers. A number of colleagues and students were actively involved in shaping the ideas presented in this book. These include David DeWitt, Haran Boral, and Shlomo Weiss at the University of Wisconsin-Madison, Michael Stonebraker and Kimman Chang at the University of California, Berkeley, and Alan Bell and Gaetano Borriello at the Xerox Palo Alto Research Center. Allene Parker provided invaluable service as proofreader, and Professor G. Schlageter of the Fernuniversität Hagen was a most encouraging editor. Of course, any remaining errors are my fault alone.

This book would not have been possible without the research support provided over the years by the National Science Foundation through grants MCS-8201860, DCR-8406123, and ECS-8403004, and the support and encouragement of Drs. W. Richards Adrion, Bernard Chern, and Lawrence Oliver.

Berkeley, CA, March 1985　　　　　　　　　　　　　　　　　　　　　　　Randy H. Katz

Table of Contents

1	**Computer-Aided Design Tools and Systems**	1
1.1	What is Design?	1
1.2	What is Computer-Aided Design?	2
1.3	Computer-Aided Design Tools	3
1.3.1	Synthesis Tools	3
1.3.2	Analysis Tools	5
1.3.3	Information Management Tools	7
1.4	The Design of Complex Artifacts	8
1.5	Failure of Current CAD Systems	10
1.6	Structure of the Book	11
2	**Survey of Engineering Design Applications**	12
2.1	Introduction	12
2.2	Basic Terms	12
2.3	Kinds of Engineering Design Applications	14
2.3.1	VLSI Design Environment	14
2.3.1.1	Multidisciplinary Design: Architecture, Logic, Layout	14
2.3.1.2	Design Methodologies: Hierarchical Approach	15
2.3.1.3	The Computing Environment for Design: Dispersed Computation	16
2.3.2	Software Engineering Environment	17
2.3.2.1	Multiple Representations: Source, Object, Runable Code	17
2.3.2.2	Design Methodology: Modular Programming	18
2.3.2.3	Configurations and Engineering Changes	19
2.3.3	Architectural/Building Design Environment	19
2.3.3.1	Pipe Design System: Sequential Execution of Applications Programs	19
2.3.3.2	Multidisciplinary Design: Piping and Structures	20
2.4	Requirements for Engineering Data Management	20
2.5	Why Commercial Databases are NOT like Design Databases	22
2.6	Previous Approaches for Design Data Management	24
3	**Design Data Structure**	26
3.1	Example: The Representation Types of a VLSI Circuit Design	26
3.2	Design Data Models	32
3.2.1	Relations (The VDD System)	32
3.2.2	A Design Data Manager (SQUID)	35
3.2.3	Complex Objects (System-R)	35
3.2.4	Abstract Data Types (Stonebraker)	39
3.2.5	Semantic Data Model (McLeod's Event Model)	40
3.3	Summary	41

4	**The Object Model**	43
4.1	Introduction	43
4.2	What are Design Objects?	43
4.3	Interfaces: How to Use a Cell Without the Details?	49
4.4	Composition and Interface	52
4.5	Complete Example of Object Specification	53
4.6	Objects Implemented as Structured Files	54
5	**Design Transaction Management**	56
5.1	Introduction	56
5.2	Design Computing Environment: Implications for Data Management	57
5.3	Conventional Transactions in the Design Environment	57
5.4	Concurrency Control Issues	60
5.5	Recovery Issues	62
5.6	Design Transaction Model	62
5.7	Extensions to the Transaction Model	64
5.8	Related Work	65
6	**Design Management System Architecture**	67
6.1	Introduction	67
6.2	System Architecture	67
6.2.1	Storage Component	69
6.2.2	Object System	70
6.2.3	Design Librarian	70
6.2.4	Recovery Subsystem	71
6.2.5	Validation Subsystem	73
6.2.6	In-Memory Databases	74
6.2.6.1	Introduction	74
6.2.6.2	Building In-Memory Structures: Complex Object Mapping	75
6.2.6.3	In-Memory Recovery	77
6.2.7	Version and Configuration Management	80
6.2.7.1	Introduction	80
6.2.7.2	Design Administration	80
6.2.8	Design Applications	83
6.2.8.1	Design Browser / Chip Assembler	83
7	**Conclusions**	84
7.1	Research Directions	84
7.2	Summary	86
8	**Annotated Bibliography**	89

List of Figures

1.1	Unintegrated vs. Integrated Design Environments	11
2.1	Alternative Types of Database Data Structures	13
2.2	Data Structure vs. Storage Structure	14
2.3	Object Dependencies	18
3.1	Block Diagram Description of Shift Cell	27
3.2	A Four Stage Shift Register	28
3.3	Waveform Representation of the Shift Cell	29
3.4	Shift Cell Floorplan	29
3.5	Shift Cell Transistor Schematic	30
3.6	Shift Cell Stick Description	31
3.7	Shift Cell as Clocked Primitive Switches	32
3.8	Shift Cell as Clocked Registers and Logic	32
3.9	Module-Parts Database Schema in System-R	36
3.10	COMP-OF Partitioning of Relations	37
3.11	Cellular Database Schema	38
4.1	Uninstantiated vs. Instantiated Objects	44
4.2	Old Shifter, New ALU within a New Datapath	45
4.3	Non-Isomorphic Representations	46
4.4	Gateways to the Design	46
4.5	Generic Object Data Structure	47
4.6	Nesting of Generic Objects	47
4.7	Designs and Libraries	48
4.8	Graphical Composition: Wiring By Interconnection	52
4.9	Design Objects as Files	55
5.1	Design Transaction Phases	63
5.2	Creating a New Design	64
6.1	System Architecture	68
6.2	Design Check-Out	71
6.3	Files Associated with a Checked Out Object	72
6.4	Internal and External Representations of a Design Specification for a NAND Gate	76
6.5	Adding an Input to a NAND Gate	79
6.6	Configuration Files and Design Versions	80
6.7	Design Life Cycle Scenario	81

1 Computer-Aided Design Tools and Systems

1.1 What is Design?

Creating large, complex systems is one of the most demanding tasks undertaken by humankind. To design a state-of-the-art microcomputer, a high performance computer system, a modern petroleum refinery, or a supersonic jetliner requires the talents of a large number of individuals (e. g., engineers, managers, technicians), coordinated into a team, making extensive use of sophisticated tools to complete the job. Much of what is designed today could not even be attempted without the pervasive use of computers. Computers automatically generate portions of the design, check the correctness of hand generated portions, and keep track of the data describing the design, especially as it evolves over time.

As engineers or computer scientists, we normally think of *design* as the activity involved with actually constructing the system; i. e., given a specification of the system, we map that specification into its physical realization (e. g., an integrated circuit chip, a computer program, a physical plant or airplane). The design task, however, extends throughout a *system life cycle,* from the initial commitment to build a new system to its final full scale production. The system life cycle includes the following activities [KATZ84]:

Initiation and Acquisition: committing to develop a new system. Example: the decsion to develop a new microcomputer system.

Requirements Analysis: determining why a new system is needed and what resources will be required for its development. Example: identification of the new market opportunities for the system, what new features it will include, and how many engineers and hours of computer time will be needed for its design and implementation.

Selection and Partition: determining whether system building blocks already exist, and chossing between alternative divisions of the system into its component pieces. Example: partitioning the system into a new processor chip and a collection of support chips (e. g., bus controllers, timers, I/O ports, etc.), some of which may already exist.

System Specification: describing what functions the system will perform. Example: choosing the processor's instruction set, and its interface to the outside world.

Architectural Design: organizing and dividing system functions across components and the people who will implement them. Example: partitioning the processor into an instruction fetch unit, execution unit, memory interface unit; assigning engineers to design and implement each of these.

Detailed Design: making the detailed design decisions for implementing each component of the system. Example: deciding on the implementation approach for the instruction unit (e. g., microcoded controller vs. random logic vs. structured logic such as a programmable logic array).

Implementation: creating a physical realization of the system components and integrating these in preparation for installation in the final system. Example: creating the integrated circuit processing masks for the processor's chip set.

Testing: verifying that the implemented system fulfills system specifications while validating its performance. Example: simulating the designed chips to verify their behavioral and electrical correctness, and to determine their expected performance.

Documentation: providing a written record of what the system does and how to use it. Example: creating a designer's notebook of important design decisions, providing documentation for the major subsystems, and writing the user manuals (e. g., "Principles of Operations") for the new system.

Use: operating the developed system under a variety of circumstances. Example: making the system available as a product, possibly to an initial set of test sites ("beta test").

Evaluation: evaluating the operation, performance, and applicability of the system in light of changing circumstances. Example: determining the commercial success of the system and tracking the bug reports that come in from the field.

Evolution: enhancing, tuning, repairing ("engineering change orders"), and converting the installed system to maintain its useful operation. Example: new releases of system microcode to provide new functions (e. g., floating point operations) or to correct bugs.

It should not be inferred from the above that a given design project will pass through each of these stages in strict sequence. Some stages may be omitted, others included, or the sequence may exhibit loops. For example, a system will be used and evaluated, suggesting avenues for further evolution. Once a new version is available, it too will be used and evaluated, and the evolution will continue.

1.2 What is Computer-Aided Design?

Computer-aided design encompasses a wide range of computer-based tools and methodologies for constructing large, complex systems. In its most general sense, computer-aided design is concerned with the use of computer-based tools to support the entire system design life cycle described above. Computer-based tools to help in requirements analysis and system specification are very much needed, but are not typically provided by existing computer-aided design systems. They concentrate on providing tools that assist the designer in physically realizing his design.

We distinguish between the terms *design automation* and *computer-aided design*. Design automation is, in a sense, more ambitious. Given an abstract specification of an object to be built, a design automation system will generate it automatically. An advantage of the automated approach is that the completed object does not need to be verified; the design automation system only generates correct objects. In the current state-of-the-art, complete synthesis algorithms exist for only a small portion of the overall design problem. Computer-aided design systems, on the other hand, are collections of tools that involve more active participation by the designer. Some of these may be automatic generators for specific portions of the design object. Other tools are explicitly directed by the designer to accomplish the design task. The difference between design automation and computer-aided design is primarily a philosophical one: are we replacing the human designer with a computer-based designer, or are we augmenting the human designer with a set of tools to enhance his productivity and to perform some of the low level detailed work. Most so-called design automation systems are actually computer-aided design systems, in that a human designer takes an active role in the design process. Complete automation of the design process remains an ambitious research topic.

Computer scientists and electrical engineers have used computer-aided design systems to construct Very Large Scale Integrated (VLSI) Circuits [CANE83], high performance computer systems [FITC82], and large software systems [WASS81] – all of great complexity. Computer-aided design tools have long been applied in other areas of expertise as well, such as mechanical system design, shipbuilding [BAND75], and architectural design [EAST81a, EAST81b]. To design larger, more complex systems in a rapid and cost effective way, designers must make even more substantial use of computer-based tools, especially to handle the vast quantity of information that describes a complex design. Structured design methodologies are needed to manage the complexity of the design task, and the tools must reflect this structure. Throughout this book, we draw our examples from VLSI circuit design, but there is much in common across design domains. Where relevant, we will point out the common and contrasting issues.

A computer-aided design system consists of three broad classes of tools: synthesis, analysis, and information management. *Synthesis tools* assist a designer in creating the object being designed. *Analysis tools* assist him in checking its correctness. *Information management tools* organize the structure of the design data, and form the foundation upon which synthesis and analysis tools can be built. The special emphasis of this book is on the latter set of tools, perhaps the most important and certainly the least understood portion of a computer-aided design system. While we do not attempt to survey the entire realm of computer-aided design tools in details (for more details, see [LOSL80, NEWT81]), each class of tools is briefly described in the next subsection.

1.3 Computer-Aided Design Tools

1.3.1 Synthesis Tools

The simplest synthesis aids are those for design capture. In fact, some tools are so simple that all they actually do is digitize a pencil and paper specification! Design capture tools enable a designer to communicate his design description to the computer. Once captured in a machine processable form, the laborious task of checking the correctness of the design – much of it manual and prone to human error – can be performed by the computer (see the next subsection). Design capture tools include language-based text editors for creating source program code, integrated circuit geometry editors for specifying the processing masks for circuit fabrication, and schematic editors for creating a description of a hardware system at the logic gate level.

A more sophisticated class of synthesis aids map a high-level description of an object into its physical implementation. The physical realization may consist of machine language instructions, fabrication mask geometries, wire-wrap lists or printed circuit board artwork, or a ship or building blueprint. For example, a programming language compiler is a "synthesis tool" that maps a high-level language program into detailed

Synthesis Aids:
* Design Capture
    e. g., Circuit Editor
* Module Generators
    e. g., PLA Generator
* Placement and Routing
    e. g., River Router
* "Design Automation"
    e. g., Silicon Compiler

machine instructions. Similarly, circuit module generators map a functional description of a subsystem, in terms of a register transfer description or input/output behavior specified as Boolean equations, into mask geometries suitable for fabrication. Circuit module generators include datapath generators [THOM83, SOUT83] and PLA (programmable logic array) generators [AYRE83].

Another class of synthesis aids are placement and routing tools. Subcomponents of the object being designed are *placed* within a two dimensional space (e. g., circuit subsystems on a silicon substrate or circuit packages on a printed circuit board) or three dimensional space (e. g., cabins of a ship). Each subcomponent has an interface, representing flows into and out of the component. These must be interconnected, or *routed*, for the components to communicate (e. g., wiring between circuits on a piece of silicon or packages on printed circuit boards, running of wiring and steam conduits between adjacent cabins of ships). A strong interaction exists between placement and routing, since a poor placement will result in a more complicated (or impossible to complete) routing problem. An initial routing is usually obtained by running the router in batch mode. For some complicated situations, the computer cannot always obtain a complete routing. The remaining routing is either completed by hand or the designer interactively assists the router in completing the task.

As mentioned above, the ultimate goal of design synthesis is to transform a very high level description of the design into a working system, essentially automating the process of design. An interesting approach is based on applied artificial intelligence research, called *expert systems* or *knowledge engineering* [HAYE83]. These systems attempt to partially automate the reasoning of an expert circuit designer in transforming a behavioral specification of his circuit into a physical realization. For example, given a collection of circuit subsystems and the needed wiring connections between them, the expert system could follow a set of rules, provided by an expert designer, in assigning particular layers of materials to interconnections (e. g., long wire runs should be assigned to low capacitance metal, metal crossing metal should use a polysilicon via, etc.). An important goal is for the system to generate several "reasonable" alternative approaches, and to allow the human designer to choose among these. The system could thus assist a novice designer in completing his circuit design [BROW83].

1.3.2 Analysis Tools

Assuming that the design has not been generated automatically, an important set of design aids are for the verification of the design. These include simulation tools, topological analysis aids, and timing analyzers.

A *simulator* is a program that models the input/output behavior of a system. The model is always somewhat abstract, and thus approximates reality rather than recreating it faithfully. Because the models can be formulated mathematically, and can be solved much more easily by computer than by hand, these constituted some of the earliest computer-based design aids.[1]

Simulation provides the designer with an alternative to building a prototype system, sometimes called a "breadboard" or a "brassboard". Simulation can be a powerful design approach when it is too expensive, too difficult, or too inaccurate to build the prototype rather than the actual system. For example, rather than simulate the design, a state-of-the-art microprocessor could be modelled by a prototype built from discrete components or TTL packages. While this makes it easier to discover and fix bugs in the logical design of the processor, the construction of the prototype is expensive, time-consuming, and the observed performance is not easily translated into corresponding performance for the single chip implementation. As more LSI components are being incorporated in high performance computers, the simulation based approach is becom-

Analysis Aids:
- * Simulators
  - e. g., Circuit Simulator
- * Topological Analysis
  - e. g., Geometric Design Rule Checker
- * Timing Analyzer
  - e. g., Program Profiler

ing widespread for large processor designers [FITC82], where prototyping had previously been the usual approach. A good description of the horror of debugging an machine prototype with logic probes and oscilloscopes can be found in [KIDD81].

---

[1] One of the first applications of computers in the early 1950's was for airfoil simulations to help design airplane wings.

Integrated circuit designers have long relied on simulation tools to help verify the correctness of their designs before their actual implementation. It is the preferred approach for debugging the design, with a strong correspondence to program testing: inputs are provided, and outputs are compared with expectations to reveal errors. Circuit fabrication involves a relatively long turnaround (usually several months), so designers cannot afford to wait for the fabricated chips before they can start to debug the design. Designers strive to obtain a working chip on "first silicon". While it is difficult to probe the internal state of an integrated circuit chip, simulations allow any internal node to be examined, greatly facilitating debugging. Because simulations allow the designer to finely control the environment in which his system is being tested, simulations are also used extensively where program testing is inherently difficult, such as in real-time applications.

The description of an object being designed exists simultaneously in several representations and at several levels of abstraction. For example, an integrated circuit design could be described by its mask geometries, its transistor schematic (where transistors are modelled either as ideal on/off switches or as analog devices), its logic gate schematic, or its functional input/output behavior (modelled in terms of Boolean equations or register transfers). Most simulation tools model the behavior of a system in only one of these representations. Along a different axis, the design's behavior could be represented at the level of abstraction of the individual transistor or in terms of transistor aggregates that correspond to functional units of the design. The time to perform a simulation is directly related to the level of detail of the representation (e. g., analog transistor descriptions versus register transfers), or its level of aggregation (e. g., individual transistors versus functional aggregates such as a register file). Simulation tools that can mix representations are called *mixed mode* simulators. For example, some circuit simulators can mix transistor and logic gate descriptions. A simulator that can compose primitive units into higher level aggregates, and simulate these aggregates faster, is called a *multi-level* simulator. A circuit simulator that composes the constituent transistors of a register file into a model of the behavior of the register file would be a multi-level simulator.

Problems in system design frequently arise from subsystem interaction. Simulators help uncover commnications errors that arise through faulty subsystem interaction in time, but provide no help when the problem is due to faulty interactions in space. Topological analysis tools are responsible for checking the correctness of the layout and interconnection of subsystems. In the integrated circuit domain, these tools include design rule checkers that insure that the mask geometries are well-formed (similar to programming language syntax checkers), connectivity checkers to insure that subsystems are properly interconnected (e. g., the flow is from outputs to inputs and not vice versa), and electrical rule checkers to insure that transistors are properly sized and matched to achieve good digital behavior. Similar kinds of aids are needed in architectural design systems. For example, interference checkers insure that the electrical wiring and water piping are not co-located, and that the structure of the building can support the pipe and wiring loads.

Timing analyzers are the final class of analysis aids. In the circuit and hardware domain, they uncover the critical timing paths in a design, providing valuable information to the designer as he tries to optimize his design. A similar function is provided in software development environments by tools that profile the running times of programs.

It is useful to distinguish between dynamic and static analysis. Topological analysis tools and timing analyzers are examples of static analyzers. They take as input a particular representation of the design, and produce as output a report describing the errors encountered, list of critical paths, etc. The analysis is independent of system inputs. Dynamic analysis tools, which include all classes of simulators, produce outputs which are critically dependent on the input set and the state of the system. The implication is that dynamic analysis tools must manage large quantities of data that are separate from representational data, i. e., the input or test data cases and the corresponding outputs.

1.3.3 Information Management Tools

The information management tools are responsible for creating, maintaining, and viewing a consistent *database* of the design description. An integrated database is the basis by which design data can be shared among the design tools of the design environment. An engineering design database has a particularly rich and complex structure, since it must organize the design data across the multiple representations of the design, its evolutionary versions, and alternative implementations. As an example, an integrated circuit design could be described simultaneously by mask geometries, interconnected transistors (transistor net list), interconnected logic gates, floorplans (the allocation of physical chip areas to particular chip functions), functional/behavioral description, block diagram description, etc. Similarly, a building design could have a structural description, an electrical wiring description, and a pipping description. The design

Information Management:
   * Design Database
   * Configuration Management
   * Release Mechanisms

database organizes the design description within each representation, correlates equivalent descriptions across the representations, and attempts to maintain these correspondences as the design incrementally evolves.

Besides issues of data organzation, managing concurrent access and update, and insuring that data is never lost due to system crashes, information management is responsible for version and configuration control. Over time, a subsystem of the design may evolve, creating an improved realization of its function. These evolutionary changes represent new *versions* of portions of the design. Putting together a particular choice of subsystem versions yields a *confirguration* of the entire system. If multiple configurations of the system are maintained at the same time, these are called *alternatives*.

To assist in keeping the design consistent, information management is also responsible for check-in procedures and release mechanisms. A new version of a design subsystem cannot be incorporated into the design description until it has been shown to be *consistent*. By this we mean that a battery of analysis tools, e. g., simulators and topological analyzers, have been applied in the proper sequence, giving high confidence that the subsystem is free of bugs and meets its design specification.

1.4 The Design of Complex Artifacts

Computer-based design tools enable us to undertake the design of systems of unprecedented complexity, far exceeding what could be accomplished by a lone designer with pencil and paper. Even with the proper tools, a complex design can easily overwhelm someone trying to understand it in its totality. This leads to errors in understanding and ultimately to implementations that do not work as expected. A number of tactics have been developed over time to reduce the complexity of a design by making it easier to understand, and we review these here (see [SEQU83] for more details):

Design Equals Documentation: In many design domains, the cost of system maintenance and evolution far exceeds the cost of the original design and implementation. This is a well known problem for software systems. Completing the design should be synonymous with completing its documentation. In large projects, the members of the design team change over time. Without up-to-date documentation, new designers cannot become productive in a short amount of time. Ultimately the entire design specification will be in machine processable form, thus making it possible to tightly couple the documentation with the most current version of the design, even as it evolves. With the documentation placed on-line, it should be made available for interactive browsing.

Use of Abstraction: Use of hierarchy and abstraction are powerful tools that humans employ to deal with complexity. Rather than attempt to understand the design of a complex system like a microprocessor all at once, it is better to view it in terms of subsystems of less complexity, e. g., as composed of a datapath and control unit. The hierarchical decomposition of the design proceeds recursively. Abstraction makes it possible to repress unnecessary details, making the design more comprehensible. For example, consider how much easier it is to understand a system when you are given its description in terms of outputs as functions of inputs rather than as a circuit layout or a transistor network! While the function can be inferred from the representation details, it is better to hide these when possible.

High-Level Descriptions: The higher the level of the description language, the more terse the description, and the easier it is to understand. A system can be developed faster and debugged easier if it is specified in terms of a high-level description language. It has long been observed that programmers write about ten lines of debugged code per day, and that engineers lay out about ten transistors per day. A good deal more work is getting done if it is ten high-level language statements or large circuit subsystems per day, rather than machine instructions or transistors.

Partitioning: A large, complex system must be partitioned into smaller, more manageable pieces in order to be understood. A functional approach is most commonly used to accomplish this partitioning. For example, a microprocessor can be functionally partitioned into (1) a control unit that interprets and sequences machine instructions, (2) an execution unit that implements the storage, and (3) a memory interface for instruction fetch and data reads and writes. Once interfaces are defined, the implementation of each unit can proceed in parallel, and the partitioning criterion can be applied recursively. Unfortunately, because of physical constraints and packaging considerations, the clean separation into functional units may have to be compromised. For example, to fit within a particular area budget, certain functionally separate subsystems may need to be combined to save on space.

Structuring Methods: Intuitively, a system is well-structured if each component implements an identifiable function, has a handful of inputs and outputs, and can be tested in isolation. The functional partitioning approach described above is an example of a structuring method. An alternative way to decompose a system is based on data or control flow. A component can be viewed as transforming its inputs to its outputs. In software systems, this is a useful way of understanding the structure of the system. However, it is particularly important in the design of physical systems, since wiring complexity and performance are critically dependent on placing in close proximity components that communicate frequently. A chip *floorplan,* which represents how the areas of the chip have been dedicated to functions, makes explicit the flow of control and data through the chip. These typically flow perpendicular to each other, so the control signals can influence the data flow.

Restriction to a Limited Set of Constructs: To reduce the complexity of a design, it is helpful to use a handful of standard building blocks and a few rules of composition. "Structured programming" advocates emphasize the use of a small number of constructs for implementing sequencing, iteration, and selection operations. Similarly, [MEAD80] describes VLSI design in terms of a small number of constructs, e. g., inverters, pass gates, nand gates, nor gates, and gives rules for their composition, e. g., a pass gate cannot be connected to the gate of another pass gate.

Testing the Design: Testing and design must proceed in tandem. Testing reveals misunderstandings between the specification of what a component is expected to do and what it has been implemented to do. The component's interface specification should include a description of its expected behavior. As each component nears completion, it should be tested in isolation, before being integrated with the rest of the system. Its observed behavior should meet its interface specification. Without a "divide and conquer" approach, testing of a monolithic system constructed from untested components is an almost impossible task.

Tools and Design Methodologies: A complex system cannot be constructed without the proper tools. The tools are most helpful when they reflect the methodology used for the design task. For example, we have discussed the importance of abstraction and the use of hierarchy in designing a system. The tools should support the hierarchical construction and verification of the system. Most existing analysis tools do not support hierarchical verification. To use them, the design must be completely specified and fully instantiated to be verified! This discourages designers from incrementally testing their portion of the design, leading to a flurry of expensive analysis activity as the design approaches its release date. As another example, the design synthesis tools can help insure that only well-formed design specifications are entered into the system. For example, an integrated circuit geometry editor can be continuously checking the design rule correctness of the layout as it is being entered by the designer.

1.5 Failure of Current CAD Systems

Computer-aided design tools must efficiently manage large volumes of information that describe a design. This includes the descriptions of the different representations of a design and versions of a design at it evolves. The information management component is crucial to the success of a CAD system, yet few systems provide sophisticated data management facilities. The current state-of-the-art is to store design data in a traditional file system. These do not provide a wide range of data management facilities, such as dynamically changeable file structures, a variety of data access mechanisms, control mechanisms for secure and concurrent access, and recoverable storage of data.

This is in part due to the way in which the systems have been assembled from individual tools. Each tool creates or analyzes a portion of the design in its own particular idiosyncratic representation. Problems arise when the designer needs to span representations. For example, to perform a simulation of the circuit described by its layout, a designer must first convert the circuit editor's description into a standard interchange format. Then he extracts the circuit into a transistor network description with a circuit extraction tool. Additional filtering programs may be needed to transform the description into the appropriate format for the simulator. Making sure that the net list representation and the layout representation continue to describe the same circuit is a difficult and error prone task[2], especially since the dependencies across the representations are not explicitly stored. Unless the tools (e. g., the editor, extractor, and simulator) use compatible representations, it will be difficult to integrate them into a design system. Even standard interchange languages do not eliminate the problem, since each tool uses "escape hatches" in the interchange language to describe idiosyncratic yet critical information. For example, the Caltech Interchange Format is the universal language used by mask-based VLSI design tools (e. g., circuit editors, design rule checkers) within the University research community, yet it has no provision for associating symbolic names with geometrical features. Different editors choose to represent the labels in different ways, making it difficult to mix editors and circuit extractors developed at different locations.

An integrated design database is a prerequisite for creating an integrated design environment (see Fig. 1.1). Placing all design information under the control of a single data management system makes it easier to maintain the self-consistency. A database is more than merely a collection of files. It *organizes* the design information across representations, alternative implementations, and evolutionary versions. By making the dependencies among parts of the design explicit, the ramifications of design changes can be more easily discovered and propagated in a controlled manner. The operational goals for the design data are to keep it consistent and durable. A data management system controls concurrent designer access and guarantees that the data will survive a system crash. These features are not always provided by a conventional file system.

Design databases have long been of interest in the computer-aided design community, and many design systems have been built on top of commercially available database

---

[2] There are many horror stories about changes made to the transistor description that were never reflected in the masks used to fabricate the chip

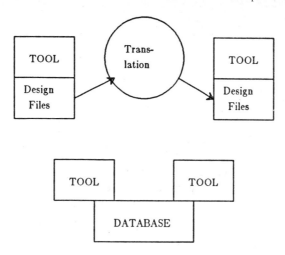

Fig. 1.1 Unintegrated vs. Integrated Design Environments. In an unintegrated environment, design data must be laboriously translated from one design tool's representation to another. A global organization must be determined for the design data if it is to be effectively shared across design tools

systems [CIAM76a, CIAM76b, FOST75, HASK82a, HASK82b, HAYN81, HOSK79, KORE75, KAWA78, MITS80, NIEN79, ROBE81, SUCH79, VALL75, WILM79, WONG79, WORK74, ZINT81]. Unfortunately, database systems have not made a major impact because of their inadequate performance and difficulty of use. Existing commercial database systems are not well-matched to the needs of engineering design applications. It is only recently that database specialists have become interested in how to best support such non-commercial applications. In the remainder of this book, we will focus on the unique data management requirements of engineering design systems, and will review the current research approaches for supporting these kinds of applications.

1.6 Structure of the Book

The remainder of this book is structured as follows. In the next chapter, we will provide a survey of engineering design applications from several different design domains (integrated circuit design, software development, architectural design). Our objective is to set out for the reader the information management requirements for engineering design systems. In Chap. 3, the variety of design representations is illustrated through a discussion of the types used in VLSI design. A number of different approaches for organizing the design database, based on actual systems, is then described. Chapter 4 contains a detailed description of one particular "design data model". A paradigm for accessing design data and for maintaining its consistency is presented in Chap. 5. Chapter 6 describes an architecture for structuring a design management system, and presents a detailed description of its components. Chapter 7 reviews the open research issues and gives some directions for future research. Chapter 8 contains an annotated bibliography.

2 Survey of Engineering Design Applications

2.1 Introduction

Before discussing the kinds of engineering data found in the design environment in the next chapter, we first review the unique systems requirements for effective engineering data management. While existing database systems provide many of the needed facilities, they by no means provide them all. Database systems have evolved to provide excellent support for commercial data processing applications, but engineering design applications are not quite the same. One of the great challenges to the database research community is to understand how existing database techniques can be applied, refined, or generalized for the engineering design environment.

First we define some basic terms, being careful to distinguish between *database management* and *design management*. We describe the VLSI, software, and architectural design environments. From these descriptions, a composite set of requirements for engineering data management will arise. With these in hand, it is easy to argue that existing database systems are inadequate for managing this data. Finally, we will review the current state of the art of appliying database technology to CAD systems.

2.2 Basic Terms

The computer-aided design and database communities often use the same terms to define different concepts. In an effort to avoid such confusion, we define the terms we will use throughout the book in this section.

Data structures are logical organizations of data. Database technologists use the term *logical data model* to denote the same concept. Database systems typically support either (1) tabular (relational), (2) tree (hierarchical), or (3) graph (network) structures among records (see Fig. 2.1). File systems can also provide data structuring capabilities. The UNIX file system [THOM78], for example, supports a constrained graph structure among directories (internal vertices) and files (leaves). A directory file can have at most one parent, but leaf files can have many parents. The internal nodes form a tree, while the leaves participate in a general graph. *Storage structures* are implementations of data structures on secondary storage (see. Fig. 2.2). These are also known as *physical storage structures* or *implementation structures*. For example, a collection of design objects organized as a tree is a data structure, while its implementation as a balanced multi-way tree on disk (e. g., a B+-tree) is a storage structure.

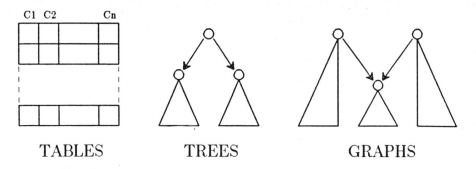

Fig. 2.1 Alternative Types of Database Data Structures. Database models typically organize data into homogeneous tables (relations), trees (hierarchies), or graphs (networks)

A *database* is the data describing the activities of an organization. Consider a VLSI chip design database. It must contain information about how design objects are composed from primitives (geometries, transistors, gates, etc.), how a design has evolved over time, who is responsible for designing its parts, and so forth. A database model provides the primitives for describing a database. The choice of which of these primitives are used to describe a specific database is called a *database schema*. For example, the names and formats of the tables used to describe a database would be its schema in the relational data model.

A *database management system* manages collections of data stored on secondary storage. It provides a standard interface to this data for application programs. The manipulation operations supported by modern systems are based on *logical* data structure rather than *physical* storage structure, so the physical structure can be changed without affecting existing programs. This feature is called *data independence,* and was not fully realized in earlier database systems. The system controls access to data in order to protect in from unauthorized or invalid access and to maintain its consistency. It provides mechanisms to insure that changes can be recovered after a system crash. *Access methods* are storage structure specific routines for manipulating data on disk. A significant distinction between database systems and file systems is the former's support for *transactions*. These are atomic actions that span multiple files, and maintain data consistency across system crashes and concurrent update by multiple users.

A *design data management system* chooses how to structure the design data within the database system. It provides a standard access interface for design tools. While the database system does not interpret the data it manages, a design management system understands how the *structure* of the data describes a design project. It enforces design data constraints. For example, part of the design data structure identifies equivalent objects across representations. Such information can help reduce the effort needed to keep the database self-consistent after a design change. The design data and the complex consistency constraints are normally what the computer-aided design community mean by "database".

A *design system* is the marriage of design tools, project management aids, and design data management facilities. The design tools create pieces of the design and validate its correctness. The project management aids assist in planning the implementation effort. Design data management is responsible for structuring the design, and for

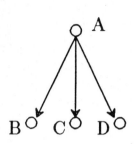
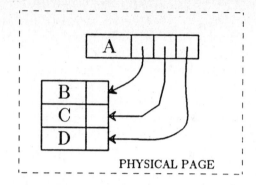

DATA STRUCTURE **STORAGE STRUCTURE**

Fig. 2.2 Date Structure vs. Storage Structure. The description of data as a tree of interrelated objects is a *data structure*. The implementation of these interrelations and objects as address pointers among records on disk is a *storage structure*

exploiting that structure to keep the design consistent. *Information management* refers to the data management activities in all aspects of the design system. Since not all of the design data need be stored in a conventional database system, we prefer to use the terms "information management" rather than "database management".

2.3 Kinds of Engineering Design Applications

While engineering design spans a large number of domains and activities, we will only focus on three representative areas: Very Large Scale Intergrated (VLSI) circuit design, software development, and architectural/building design. These will serve to motivate the requirements for engineering data management.

2.3.1 VLSI Design Environment

2.3.1.1 Multidisciplinary Design: Architecture, Logic, Layout

A VLSI design is specified in a number of equivalent, but not necessarily isomorphic, representations. By this we mean that the precise decomposition of a system into components and subcomponents is not identical across all representations. However, each description is complete, but represents the system from a different viewpoint. The details of the language for specifying the design depend on the representation. For example, the chip system architects are responsible for the register transfer description, which specifies the functional behavior of the chip. The logic designers create the logic description, usually specified as logic gate schematics and drawn on a CRT display. The ultimate manufacturing specification, the layout masks, are created by the layout artists on color displays. Due to the desire to get the product to market as soon as possible, the representations are often refined simultaneously, rather than sequentially.

A common theme that emerges in design activities is the need for three different "dimensions" of the design. The architectural view is representative of the *functional* dimension, i. e., a specification of the behavior and interactions of pieces of the design. The logical view is along the *schematic* dimension, wherein the types of design parts and their interconnections are represented. The layout geometries illustrate the *physical* dimension, in which the shapes and spatial representations of design parts are emphasized.

Keeping the design description consistent within and across representations is a critical challenge. The layout representation is checked for correctness by the design rule checker, which insures that the geometries overlap and touch in allowable configurations. Depending on the style of the logical description, this representation can be checked by a logic simulator (if the specification is in terms of logic functions), a switch level simulator (if the transistors are modelled as idealized switches), or a circuit simulator (where the transistors are modelled as analog devices). The register transfer specification can also be checked by simulation.

Besides consistency checking within a representation, we must also check for consistency across representations, e. g., to verify that the layout description corresponds to the associated logical description. The circuit description is extracted from the layout, during which electrical parametes are also extracted and connectivity/electrical checking is performed. The latter checks that all transistors are properly sized to obtain good digital behavior, that no portions of the circuit are unconnected, and that certain "sanity" conditions hold. e. g., power and ground have not been wired together. The extracted representation is suitable for simulation. Equivalence between it and the logical representation is shown by either direct schematic comparison (are the interconnected networks of transistors equivalent?) or by simulating each in tandem and comparing the results (is their output behavior identical for the same inputs?).

2.3.1.2 Design Methodologies: Hierarchical Approach

A hierarchical design approach is used almost universally. The design description is structured as a forest of cell hierachies, one hierarchy per representation domain (e. g., layout, logic, functional). A cell is an identifiable unit of the design: a transistor or gate, or an aggregation of these, such as an adder, a register, a datapath. The root of each hierarchy represents a different view of the object being designed. If it is a microprocessor, then the root of the layout hierarchy represents the layout for the complete processor, while the root of the logical hierarchy represents its complete logical description. The children of the root represent the system's decomposition into major subsystems, such as the datapath and control unit of the processor. These subsystems are further decomposed into more detailed subsystems at the next lower level, and so on.

A hierarchical design approach does not necessarily imply a top-down methodology. A designer begins by either describing his system with a block diagram or by providing a procedural description of its function. Either description is expressed in terms of subsystems, which themselves need to be designed or extracted from a library. He then proceeds to describe each subsystem – and its interaction and interconnection with other subsystems – as he further refines the design. This approach is normally taken when a full custom design is being undertaken. A designer may also piece together a design from a library of building blocks, proceeding in a bottom-up fashion. This approach is often taken with gate array and standard cell methodologies. In either case, the designer will need to span representations, giving more breadth to the design as he proceeds.

Organizing the design into a hierarchy supports a conceptual partitioning of the design, which is crucial for supporting design teams. A designer is responsible for a conceptual unit of the design, corresponding to a subtree of the design description. While we consider trees for simplicity, rooted directed acyclic graphs are easily accommodated. Within the subtree, the designer can proceed as he wishes, as long as his implementation does not alter the specified interface and function of his portion of the design. Alternative implementations of the same portion of the design can be represented as subtrees in parallel with the root of the original subsystem.

2.3.1.3 The Computing Environment for Design: Dispersed Computation

The environment for VLSI design is rapidly emerging as one in which dedicated designer workstations are attached to service machines via a local network. The workstations typically contain a powerful microprocessor, high quality graphics display (usually in color), a sophisticated pointing device (e. g., a mouse), and an optional hard disk. The file servers can be more conventional mainframes, with multiple disk controllers and drives, although they are frequently identical to the workstations. At least in the short term, the environment is characterized by a high degree of heterogeneity in both software (design tools acquired from different vendors or mixed with tools developed by different groups in house) and hardware (general purpose mainframe computers, general purpose workstations, graphical entry stations such as CALMA and APPLICON). Just making these disparate elements of the design environment communicate is a major effort.

The process by which portions of the design are created is both *tentative* and *iterative*. By tentative, we mean that the activity of design is exploratory: a designer tries out a number of alternative approaches to the problem, some of which may turn out to be dead ends. By iterative, we mean that a design proceeds through a number of evolutionary versions: a design is refined piece by piece and at each step, more details are added, corrections are made, and enhancements are incorporated.

In such a computing environment, a reasonable model of how a designer interacts with his data is as follows. He first checks-out a portion from the public file server copy of the design into a private workspace. Over time, the design portion is modified by the designer with the tools available to him at the workstation. When he is satisfied that his changes are complete, he checks the design portion back onto the file server. However, before it can be incorporated into the public copy of the design, it must pass through a battery of consistency tests, such as simulations, design rule checks, etc. Only self-consistent design portions should be made publically available.

The design process is iterative, and a design will evolve over time. A designer will try alternative design approaches or will make improvements to an existing implementation. It is desirable that a new tentative version NOT overwrite an existing public version. Once a design portion is checked out, it is returned to the public repository as a new version. For example, when the design activity is complete, a complete chip description is assembled from a particular configuration of versions of its parts, and is released to manufacturing.

Because of the iterative and tentative nature of design activity, the work done in modifying the design at the workstations resembles a text editor session. A designer must be able to save in-progress changes (i. e., make these survivable across system crashes), leave parts of the design incomplete and unspecified, and be able to undo some of the changes, perhaps back to a previous savepoint.

The notion of *design check-in* is intimately related to version control. Old versions of data are never overwritten. Check-in creates a new version, which is added to those available on-line. Strategies for archiving very old versions, by migrating them to off-line storage, are clearly needed.

Because only *consistent* versions of design parts can become visible to other designers, the system enforces a stricter control over the design than that which is normally provided by simple version control mechanisms. Note the close interaction between the verification of the design and the creation of its new version. This has been called the *system release cycle*.

2.3.2 Software Engineering Environment

2.3.2.1 Multiple Representations: Source, Object, Runable Code

There is much in common with the VLSI design environment described above. In fact, VLSI design incorporates many aspects of program design! Many VLSI processors are implemented with microcoded control stores. The functional description of the system may be specified in terms of a high-level hardware description language. Some researchers advocate that the layout be specified in a language for generating geometries, rather than creating these directly with a layout editor. Thus, even the VLSI design environment needs support for good software programming practises.

A programmer's primary "design capture tool" is a text editor. Some of these have been specially designed for particular programming languages, and are called *language directed editors*. Syntactic and semantic checking can be performed while the program text is being entered. Compilers are the "synthesis aids" that map programs written in a high level language into runable machine code. While there appears to be a single design representation from which the others can be derived – namely, the source code – this is not true! A number of pieces of information about the design, such as the documentation, the requirements specification, project schedule, and test data cases, cannot be algorithmically derived from the source code.

Programs are usually tested by running them in their applications environment with controlled test cases. *Debuggers* permit the programmer to examine and modify the run-time environment while testing the program. The program test data is an important part of the overall design database.

The object code representation is derived from the source code, and the executable code from the object code. To keep these representations of the program consistent, dependency analysis and enforcement tools have been developed, such as the UNIX MAKE facility [FELD79]. Consider the case where many program modules depend on a shared data structure. When this is changed, the modules that refer to it must be recompiled, and the system must be reassembled. Dependency analysis tools allow the programmer to specify such dependencies, and to give a script of commands for what to do when the dependency has been violated. For example, if module Y depends on module X (it must be newer than X), and X was last updated after Y, then a script for regenerating Y can be invoked automatically. This may spawn further regenerations for modules that depend on Y (see figure 2.3).

2.3.2.2 Design Methodology: Modular Programming

Modern programming languages encourage a style in which a system is created from a large number of small modules. Extensive use is made of library modules, providing standard utility functions such as string manipulation and input/output. A major information management issue is how to keep track of the system's pieces as they undergo constant modification and improvement.

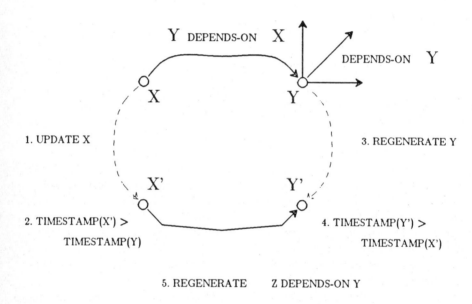

Fig. 2.3 Object Dependencies. Y depends on X. If X is updated, then its timestamp is newer than Y, and a new version of Y must be generated. After regeneration, Y's timestamp is once again newer than X's. This change may spawn additional regenerations for those objects that depend on Y

A collection of techniques called *source code control* have evolved for maintaining control over the evolution of a program. Program text is maintained as versions; old versions are not overwritten. It is not unusual for the debugging process to diverge, i. e., fixes to some bugs may introduce new bugs. Therefore, it is useful to be able to "roll back" program text to an earlier point known to have passed certain acceptance tests. In addition, the source code control system allows changes to be annotated: who made the change, when, and for what reason, providing an audit capability.

2.3.2.3 Configurations and Engineering Changes

Besides changes to the source code, the *configuration* of the system can change over time. A configuration is a collection of the versions of particular modules from which the system is composed. At different points in time, subsystems of the program may be implemented by a different composition of modules. A configuration control system is responsible for maintaining this composition information.

Certain distinguished configurations are called *releases*. These represent major versions of the system that are usually made available to users outside the design group. When approaching the creation of a new release, all versions of modules used in its creation are frozen. Extensive testing of the system is performed, and discovered bugs are corrected very carefully, least they introduce new bugs. At some point, the system's developers are satisfied with the functionality and correctness of the release, and ship it to users. In the meantime, system development continues and bug reports from the field are either fixed or replaced by new code in the next release.

2.3.3 Architectural/Building Design Environment

2.3.3.1 Pipe Design System: Sequential Execution of Applications Programs

The following set of applications is adapted from [EAST81b]. The programs synthesize the piping system of a building or processing plant, taking into account flow requirements to determine the pipe sizing and checking for conflicts with other representations of the design. Unlike VLSI or software design, the design of piping systems is a much better understood process (such things have been being designed for hundreds of years!), and has been more successfully automated than the other design domains we have examined. There are aspects of these systems that make them extremely complicated, however: the design system must deal with design components that interact in 3-space. VLSI and software are more "two-dimensional" in nature. Piping system design is representative of a variety of physical CAD/CAM applications in architectural, shipbuilding, and aircraft design.

The piping application consists of (1) piping system topology design, in which a piping network is interactively described as a set of source and destination nodes, with their physical locations; (2) pipe sizing design, in which flow requirements and the material transported determine the size of the pipe and material from which it is constructed; (3) pipe fitting and shop drawings, in which the detailed fitting of the piping network is determined and the detailed shop drawings are generated; (4) cost analysis, in which the pipe and fitting costs are estimated; and (5) machine control generation, for the generation of machine control tapes for pipe cutting, bending, welding, and threading.

Each program (except the first) takes as input the outputs from the previous program and also accesses an application specific database. For example, the pipe sizing program makes use of a database of material properties; the pipe fitting application accesses a database of standard fittings; the cost analysis program uses a database of cost data.

Although the programs are run sequentially, the design can progress iteratively. At any step in the application sequence, we can retrace to a previous step, modify the inputs, and iterate the sequence that follows. A designer might do this after the cost analysis, discovering that the described piping network is too expensive to implement. Since the application consists of sequences of program executions, dependencies exist among outputs of some programs and the inputs of others. As in the software environment, these dependencies must be maintained to keep the application results consistent.

2.3.3.2 Multidisciplinary Design: Piping and Structures

The piping application depends on other aspects of the building design process. For example, the building structure must be able to support the pipe, and structural elements cannot be co-located with pipe elements ("spatial conflicts"). These introduce further constraints on what constitutes a valid piping network.

The structural development application consists of (1) generation of the structural network (i. e., the structural frame of the building); (2) analysis of the structure, to determine properties of its sections; (3) determination and sizing of support members, including a cost estimation ; (4) detailed design of support structure joints; and (5) production of shop drawings. The interaction of the piping and structure subsystems occur in step 3. The member sizing must take into account the pipe loads it will be called upon to support, in addition to the structural loads themselves.

2.4 Requirements for Engineering Data Management

A number of reoccurring themes are evident from our descriptions of three applications environments. These include hierarchical design approaches, support for multiple design representations, consistency constraints that span these, version and configuration control, and automatic regeneration of portions of the design when a portion has changed. We examine these themes in more detail in this section.

Requirements for Engineering Data Management
* Hierarchical Data Structure
* Multiple, Correlated Design Representations
* Cross Representation Consistency Checking
* Version Control
* Programming Language Interfaces
* Integration with Graphics and Drafting
* Structured Entity Manipulation
* Conventional Database Manipulation
* Dispersed and Distributed Data Processing Support
* Integration with Manufacturing and Project Management

Hierarchical "levels of abstraction" are important for tackling any complex problem (as we have seen in Chap. 1). VLS systems, software systems, and even piping systems can all be conceptualized as a complete design first, and then recursively decomposed into constituent, more primitive pieces. The approach is well-known in software design, and has greatly influenced design methodologies for VLSI. Even in piping system design, the architect thinks of the system first in terms of a whole plant, divides these into zones, then further divides these into floors, and finally rooms. The leaves of the hierarchy represent the detailed specification of the primitive cells that constitute the design. Internal nodes describe how the more primitive components are composed and interconnected. To adequately support design activities, the design tools must reflect this hierarchical structure. The design data management system is no exception.

While a hierarchical approach is generally agreed upon, the details of design representation vary from domain to domain, and even within a given domain, controversy exists. A VLSI design can be represented as layout geometries, stick diagrams, block diagrams, logic diagrams, transistor diagrams, functional specifications, or behavioral specifications. Software is represented as source, object, and binary codes, as well as by documentation and statements in a requirements specification language. In the architectural domain, the landscape architect has one view of the design, the structural engineer another, the electrician another, and the pipefitter yet another. A design management system must be able to support a variety of design representations. New representations appropriate to new design methodologies will need to be supported for evolving domains such as VLSI design; thus, a priori assumptions about the kinds of supported representations cannot be imposed. A design data management system should provide mechanisms for organizing a design description across multiple representations, especially capturing the dependencies across these.

Mechanisms for keeping the design consistent are also needed. While it may be difficult for a design management system to automatically propagate changes across representations (this requires the detailed semantics of the design representations to be interpreted by the system), it should at least use the structure of the design database to identify the possible ramifications of a design change. For example, if a portion of the design is changed in one representation, the system should flag the portions of the design in its other representations that could have been affected by the change.

Keeping track of and organizing versions of design objects is an important task of a design manager. Efficient access to versioned data, stored with a minimal amount of physical redundancy, should be supported. Facilities for declaring a design "released" and no longer subject to update, along with mechanisms insuring that the appropriate approval, sign-off, or check-in procedure has been followed, is also needed. History tracking aids, and the ability to perform a comparative analysis of versions, are additional needed features.

Since some design tools currently exist as batch applications programs, an interface to the database from standard programming languages must be supported. Since many existing design applications are coded in FORTRAN, e. g., circuit simulators, structural analysis programs, etc., a FORTRAN interface to the engineering database is a necessity. In some design domains, e. g., aircraft design and circuit simulaton, scientific data types should be supported – for example, matrices of real or complex numbers. Because of the spatial nature of many design activities, e. g., pipe system layout and VLSI mask artwork, support for geometric data types is needed. Because such data is most naturally created and viewed graphically, e. g., as a blueprint or a mask artwork checkplot, the database must be interfaced to design drafting and graphic display systems.

Engineering applications manipulate structured entities, e. g., a collection of VLSI geometries oriented in a two-dimensional space that constitutes an adder, a collection of source code text that implements a string package, the descriptive information about a pipe as it routes through a processing plant. These objects are stored as interrelated collections of records, rather than as individual records, and are manipulated as a logical group. They are the natural units of creation, access, and manipulation, and represent partitions of the database that are relevant for particular engineering tasks. A design management system should provide operations for manipulating such objects.

While engineering applications often access their design data organized as complex objects, they continue to access data structured in more conventional ways, such as part catalogs, inventories, and accounting data. The data management system must continue to support conventional database interactions, as well as supporting interactive design.

The design management system must provide access to design data in a particularly complex computing environment. The environment contains aspects of dispersed computing, such as designer workstations spread around a design office. Because design and manufacturing activities frequently take place in different locations, and since these share much common data, the design database is also geographically distributed. The current design environment also exhibits a high degree of heterogeneity, both in terms of software and machines. This leads to even greater communications problems.

While we tend to focus on design, we must not forget the information management needs of the manufacturing process. Manufacturing applications include shop scheduling, materials requirements planning, inventory control of finished goods, and accounting. Conventional database systems have evolved to support just these kinds of applications. In addition, however, support is needed for storing the manufacturing specification, e. g., machine tool control tapes, for controlling a step and repeat camera for VLSI mask making, or for the tooling of pipes.

Finally, special information must be stored for project management purposes. It should be possible to generate the management reports describing whether the project is on schedule, whether it is within budget, whether critical portions are being completed on time, etc.

2.5 Why Commercial Databases are NOT like Design Databases

Database facilities have evolved to support both high performance transaction processing and interactive use by non-programmers (an excellent description of commercial database system technology can be found in [DATE81, DATE82]). These include (1) structures for efficient access to data on secondary storage, (2) the concept of a transaction: collections of read and update operations treated as atomic units of database conistency, (3) protection and concurrency control mechanisms for controlled data sharing, (4) automatic integrity maintenance, (5) crash recovery services, and (6) user-friendly interfaces (graphical, natural languages) and high-level query languages.

Not all of these services are quite what is needed for design systems. Consider the need for efficient access to secondary storage. While fast access to records is important, the overhead of entering and leaving the database system to extract a record at a time is too great for the large quantities of data involved. It is more efficient to extract and replace large aggregations of design data as a unit.

The transaction concept has been one of the crowning achievements of database research [GRAY81]. Transactions are sequences of database read and write actions that leave the database consistent: interleaved transaction executions are permitted as long as their result is the same as some serial execution of the transactions. Transactions are atomic: all changes become visible at once (the transaction *commits*), or none become visible (the transaction *aborts*). Transactions are durable: once a transaction has committed, its changes are permanently installed in the database even if the data should be temporarily lost because of system failure.

Transactions seem like an excellent mechanism for assisting in maintaining the consistency of a design database. However, conventional transactions have been developed for the short duration/simple units of work typically found in transaction processing environments, such as airlines reservations. Unfortunately, they do not adequately model design interactions. A *design transaction* begins with a designer acquiring exclusive access to a portion of the design, modifying it over a long period of time, and committing the changes only when they have been shown to be valid.

Design interactions are not atomic: they more closely resemble editor sessions. Normally, a database system recovers from a crash by undoing the effects of incomplete transactions and redoing the effects of completed transactions. If the duration of a transaction is from the time an object is checked out from the repository to when it is returned, then hours or days of work could be lost if the system crashes and the conventional strategy is followed. The database should be returned to the most recent state possible, even past the last state "saved" by the designer.

Design interactions require a stricter notion of consistency. Although database systems can automatically support integrity constraints, they are of a very simple form only. Complicated constraints on design data are not easy to specify or enforce. Compare the constraint "salary must be greater than zero" with "the circuit must behave as specified with expected performance".

Design interactions are even more durable than conventional interactions. Design data lives beyond subsequent transactions. Design transactions create new versions of the objects they update. Conventional databases are optimized for maintaining the current version of the data, and database systems provide no support for versions. Of course, conventional mechanisms for recoverable update and archive must be used to insure that old versions are never lost because of a system crash.

Many database implementation issues are actually simpler in the design environment. Because of design teams and their strict partitioning of tasks, designer interference is rare. While many sophisticated techniques have been proposed for controlling concurrent access, simple techniques are sufficient to resolve most conflicts for design applications. Data sharing can be controlled by not permitting moe than one designer to update the same portion of the design at the same time. If this proves to be too restrictive, designers can negotiate among themselves to determine who gets to update the object next. Versions greatly simplify the implementation of concurrency control and recovery mechanisms. Further, the complicated query processing algorithms may not be needed. It is more natural to look for a design object of interest by traversing through a design database in an interactive "browsing" manner than to formulate a query in a high-level query language.

Database systems do not adequately handle the basic storage needs of design data. They have been tuned for large volumes of regularly structured data. Design data is organized into complex structures, with large numbers of interconnected files of relatively small size. It is not easily formatted for storage in a conventional database. Few systems adequately support the kinds of variable length heterogeneous data typically created by design tools.

As can be seen from the above, there is an applications mismatch between what design tools need and what database systems provide. In the remainder of this book, we shall describe the structure of a system that can satisfy the data management needs of design tools.

2.6 Previous Approaches for Design Data Management

A major weakness of earlier systems is their inflexible choice of the set of supported representations. While this may be acceptable for well-understood design domains, such as building design, it is not the case for evolving design areas. For example, the correct set of representations for describing a VLSI design has yet to be determined.

A number of design systems have been founded upon a database approach. Each is lacking because it has been built around a predefined database structure. As examples, [TAKA80] chooses to represent a design as a network of devices. [ZINT81] describes a design in terms of a logic diagram in addition to the network. [SUCH79] simply describes a design as an interconnection of parts from a standard catalog of discrete transistors, SSI, MSI, and LSI parts. [CIAM76a, CIAM76b) represent a design as devices, pins, interconnections, and signals. [KORE75] describes a design in terms of parts, nets, wiring, and layout artwork. Logical, physical, and electrical representations are part of the database described in [WORK74]. Unfortunately, none of these systems make it easy to create new design representations.

In addition, none of the systems mentioned above have been implemented with state-of-the-art database systems. All have been built on earlier generation systems that support the network model of data. These provide primitive access and concurrency control features, and do not adequatly isolate the design management software from the physical data organization.

Some systems have been based on the more recent relational techniques [WILM79, VALL75]. However, these systems do not address the issues of organizing a design hierarchy or supporting multiple design representations.

[ROBE81] describes a system that comes close to being complete. It supports multiple design representations constructed from a set of basic data organzation (symbol tables, relational tables, structured data tables, and unstructured data) rather than predetermined design object types. Interfaces are provided for a variety of tools. However, it fails to support a design hierarchy, and provides no mechanisms for maintaining consistency across representations. Further, it is built on top of a CODASYL database system, implying that it has poor facilities for access and concurrency control, and is overly dependent on the physical organization of the database. Another weakness is that a CODASYL database schema is static, and thus, new design representations cannot be added without a major reorganization of the database.

The GLIDE System [EAST80] is an excellent example of recent work in design data management. Its features include dynamic schemas, automatic invocation of procedures for integrity enforcement, and record types and operations for geometric modelling. Further, the system is accessed through its own sophisticated programming language, rather than by a procedure call interface. However, because it is oriented towards general purpose solid geometric modelling, its features are not appropriate for all design domains. In particular, a design hierarchy is not explicitly supported, and thus cannot be used to aid in design change propagation or to control concurrent access. A flexible choice of design representations is not supported either, although there is some support for design alternatives. Finally, the choice of interface, i. e., through a special programming language, may make it diffcult to interface existing tools to the design system.

The utility of database techniques for design data management has not been universally accepted by the design automation community. [SIDL80] makes the claim that commercial database systems are both difficult to use and suffer from poor performance. [BAND75] claims that certain useful operations, such as tree traversal, are difficult to perform in a relational system. However, this is more an artifact of high-level relational query languages than the tabular data organization. A system with a low-level interface could support the operation. [TRIM81] claims that extensions to a database are disruptive and costly, and that interfacing to a database requires too much work. The first claim is not as true for relational systems, which support schemas and structures that change dynamically. The latter issue is addressed by the remainder of this book: how can a system be designed to make it easier to interface design tools to a database.

No existing, commercially available system supports the complete range of facilities needed to support design activities. in particular, the missing features include: an explicit representation of the design hierarchy, support for a flexible choice of design representations, and a multi-level architecture, providing a design object oriented interface to design tools on the one hand, and an efficient mapping onto physical disks on the other.

3 Design Data Structure

In this chapter, we will examine different ways of describing the *representational* and *structural* details of a design. Since the representational details are typically domain specific, we will limit the discussion to one design domain, namely VLSI design. We then describe several proposed and implemented *design data models*. These provide structures for organizing the design description within and across representations, for incorporating interface specifications, and for representing design versions and alternatives. A more detailed description of one such model, the *Object Data Model,* will be presented in the next chapter.

3.1 Example: The Representation Types of a VLSI Circuit Design

VLSI circuits are a particularly interesting design domain because of the rich set of representation types used in their description. The large number of types is due to a number of reasons: the continuously changing design methodologies attempting to keep pace with evolutions in technology; the need to simultaneously span topological, electrical, and logical concerns in order to make an effective design, and thus, the need to resort to various "mixed representations"; the lack of widespread standards for all but the lowest levels of design description (e. g., the physical layout).

We will use a very simple VLSI subsystem to illustrate the kinds of representations presented below. Consider the design of a simple shift register cell. The behavior of the cell is most easily understood if we first understand its input/output behavior in terms of the system's timing. The subsystem is clocked with a two phase (PHI1/PHI2) non-overlapping clock. It has one input signal IN and one output signal OUT. When the first clock phase is asserted, the input is sampled and latched internally. During the second phase, the new internal state is revealed on the output line. The output remains valid through the next PHI1, even though the input may have changed, thus changing the latched state.[3] The output does not change until the following PHI2. VLSI building blocks often exhibit this ability to hold two states in the middle of a clocking sequence. A ful PHI1/PHI2 cycle is required to propagate the input change to the output.

Now, we shall describe the ways in which such a subsystem might be represented in a VLSI chip design database. We also indicate the analysis tools typically used to verify the description's correctness.

---

[3] For those readers familiar with conventional digital design, but not VLSI design, such a shift element can be implemented by a master/slave flip flop

(1) **Block Diagram:** Circuit subsystems are represented as named boxes, with input/output signals to denote control and data flow. Signal busses, i. e., wires that connect more than two subsystems together, are also specified. This representation is used primarily as a documentation and organizational aid.

The shift cell is represented as a black box with input signal IN, output signal OUT, and two control signals, PHI1 and PHI2, to represent the clocks (see Fig. 3.1). While not very interesting for such a simple subsystem, the block diagram representation is more useful for describing larger aggregations of the design. The key information it presents is how subsystems are *wired* together, i. e., how the outputs of one subsystem provide the inputs to another. For example, the block diagram description of a four stage shift register would appear as shown in Fig. 3.2.

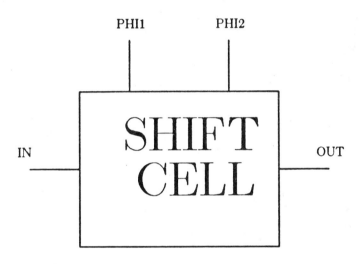

Fig. 3.1 Block Diagram Description of Shift Cell. A design object is described as a black box with interconnections

(2) **Behavioral/Functional:** The behavioral representation is similar to the block diagram representation, except that the output signals are defined as functions of the input signals. This description is most frequently used to partition the design among subsystem designers. Functional simulators are used to verify the correctness of a behavioral representation.

Note that there are many different forms for the behavioral description, depending on the subsystem being described. A cell consisting of combinational logic only (i. e., no internal state caused by feedback circuits) can be described as a *truth table* of input combinations for each output signal. For subsystems with state, the outputs are a function of the current state as well as the inputs, and the behavior can be described in terms of *state diagrams, state transition tables,* or even *programs* written in a hardware description language.

28   Chapter 3. Design Data Structure

For example, our simple subsystem could be described as follows:

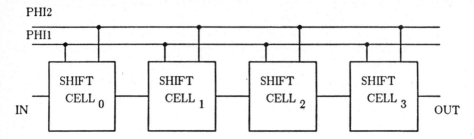

Fig. 3.2 Four Stage Shift Register. Block diagram description of a four stage shift register. The interconnectivity and busing structure of the design is made explicit

```
LOCAL STATE InLatch, OutLatch;
INPUTS In; OUTPUTS Out;
CLOCKS Phi1, Phi2;

IF Phi1 THEN InLatch <- In,
IF Phi2 THEN OutLatch <- InLatch,
Out == OutLatch;
```

This small fragment illustrates some of the issues in hardware description languages: what is the internal state (InLatch and OutLatch), what is the system clock (Phi1; and Phi2, which are implicitly non-overlapping), and under what conditions does the state change (namely, when the clocks are asserted). InLatch and OutLatch are internal state registers. If Phi1 is asserted, inLatch changes its state to whatever is on the IN signal. On Phi2, OutLatch changes its state to the value of InLatch. OUT is defined as always being equivalent to the value of OutLatch. Note that if performance information is associated with the behavioral description – for example, to specify how long it takes for the InLatch or OutLatch to change state and propagate signals – then a *waveform* representation may be more appropriate. For example, see Fig. 3.3.

(3) **Geometrical/Physical:** The physical representation describes how pieces of a design are physically placed. The design can be viewed as a collection of tiles that are placed adjacent to each other in a two dimensional space. If the tiles represent the materials from which the circuit will be constructed, e. g., units of polysilicon, diffusion, or metal, then the description is called the *layout* or *mask geometry* representation. If the tiles are similar to block diagrams, except that they have topologically accurate placements of the input/output connection points, then the representation is often called the *floorplan*. Tools that manipulate the physical representation include those for placement and routing (used in composition cells), cell stretchers and compactors (used to modify leaf cells), and synthesis tools that generate this design representation from the behavioral, geometrical, or sticks representations. Design rule checkers check the validity of the geometries.

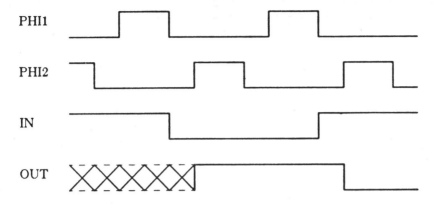

Fig. 3.3 Waveform Representation of the Shift Cell. The behavior of a design is specified as corresponding waveforms for inputs and expected outputs. Such a specification is particularly useful when simulating or testing a design

For such a simple subsystem as the example, the floorplan need not be very different from the block diagram view. The difference is that ordering of signals around the periphery of the block is significant, as these will map directly to the locations of the signals in the implementation. Some important signals, namely power and ground, are usually left out of the block diagram description but are included int he floorplan view (see Fig. 3.4).

(4) **Electrical/Transistor:** The electrical representation describes a design as an electrical circuit, usually as a circuit schematic (i. e., interconnected transistors with associated resistances and capacitances). The verification tools my be at the analog level (transistors modelled as analog devices), or at the switch level (transistors modelled as perfect switches). The representation is appropriate for detailed timing simulations. Electrical rules checkers validate the electrical representation.

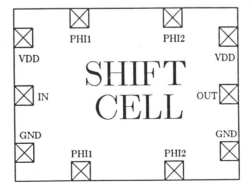

Fig 3.4 Shift Cell Floorplan. The floorplan is similar to the block diagram, except that the placement of interconnection "ports" corresponds to where those interconnections would be made in the physical layout of the design

30 Chapter 3. Design Data Structure

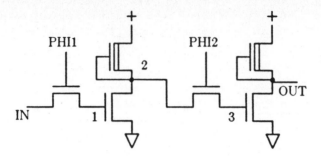

Fig. 3.5 Shift Cell Transistor Schematic. The transistor schematic specifies the interconnection of the transistors constituting the electrical design. The electrical nodes are labelled 1, 2, 3, IN, OUT, PHI1, PHI2, Power ("+"), and Ground (inverted triangle)

The example can be implemented by a collection of six MOS transistors (see. Fig. 3.5). Each of these devices has a drain, a gate, and a source terminal.[4] Without getting into the details of how the device operates, it is enough to think of it as a voltage controlled switch. When a logical "1" voltage is applied to a transistor gate, then the source and drain are connected. When a "0" voltage is placed on the gate, the source and drain are disconnected. The important information represented here is the interconnection of the transistors, i. e., which transistor source/drain is connected to which other transistor's gate, source, or drain. While Fig. 3.5 shows a *schematic* view of the subsystem, another possible representation of the same information is the *netlist*. This consists of listings of devices and the assignment of node numbers to the terminals of each device. If two devices share the same node number, then they are connected. For example, the shift cell netlist is the following:

(Transistor-type[5]   Drain-node   Gate-node   Source-node)

(E   IN   PHI1   1)
(E   2   1   GND)
(D   VDD   2   2)
(E   2   PHI2   3)
(E   OUT   3   GND)
(D   VDD   OUT   OUT)

In addition, important physical information, such as the areas and perimeters of the gate, source, and drain are also recorded. The connectivity information is used by the switch level simulators, while performance simulators (e. g., SPICE) and

---

[4] A fourth terminal, the substrate, is also included in the circuit description for electrical completeness, although it is not necessary for logical simulation of the device

[5] E = Enhancement mode device; D = Depletion mode device.

timing analyzers (e. g., CRYSTAL, TV) use both connectivity and physical data to determine the timing behavior of the circuit.

(5) **Sticks:** This representation combines the topological properties of geometries with transistor switches. A sticks description is easy to stretch and compact, making it a higher level description from which the geometries of the physical layout can be synthesized. It is validated by switch level simulation.

The circuit level description given above does not include the critical information of where the interconnected transistors are placed with respect to one another. STICKS provide the information, without cluttering up the representation with actually drawn devices. A transistor is represented by one kind of wire crossing another. The stick description of the shift cell is given in Fig. 3.6.

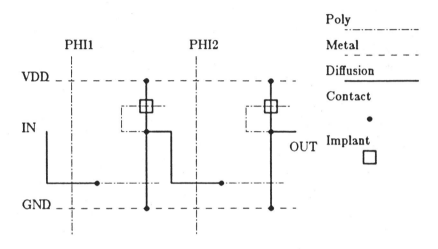

Fig. 3.6 Shift Cell Stick Description. The sticks description combines the "logical description" of transistor schematics with the topological description of layouts

(6) **Clocked primitive Switches/Logical:** At the logical level, most conventional SSI or MSI designs can be described in terms of interconnected logic gates. Unfortunately, VLSI circuits can rarely be described completely in this way, because of the use of circuit structures that have no logic gate equivalent. [BELL81, STEF82] propose a new mixed description called Clocked Primitive Switches (CPSW). The representation combines concepts from the electrical level and logical level to insure that circuits behave correctly as logic devices. Eletrical/logical checkers can verify the correctness of this representation.

For example, the configuration of two transistors that perform the logic function of negation can be aggregated into a single inverter. The shift cell can be represented NOT as six interconnected transistors, but as two *pass gates* and two *inverters* interconnected as in Fig. 3.7. The pass gate is a special "charge directing" structure that cannot be easily handled by simulators built for conventional NAND, NOR, and NOT logic. From the data representation standpoint, however, the information is similar to that of interconnected transistors, except that at least some of these have been aggregated and replaced by the logic gate they implement.

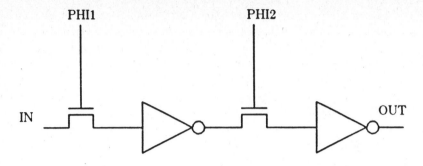

Fig. 3.7 Shift Cell as Clocked Primitive Switches. The design is represented as gate logic interconnected by alternating stage of special charge directing logic

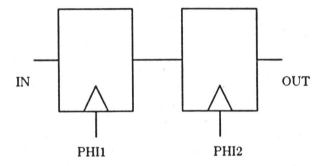

Fig. 3.8 Shift Cell as Clocked Registers and Logic. The design is represented as two-phased clocked registers interconnected with unclocked combinational logic

(7) **Clocked Registers and Logic:** This is another hybrid representation suggested in [BELL81, STEF82] that is similar to the behavioral level, but explicitly includes knowledge about the clocking of circuit elements. A variant of a functional simulator can be used to validate this representation.

The shift cell would be represented by two interconnected registers: one clocked on PHI1, and the other clocked on PHI2. The pass gate and inverter of the previous representation type have been aggregated into a *register*. Thus the primitives of the representation are interconnected blocks of registers and combinational logic. In the case of the shift cell, there is no logic other than that used in implementing the register (see. Fig. 3.8).

3.2 Design Data Models

3.2.1 Relations (The VDD System)

The VDD System [CHU83] is an integrated VLSI design system, developed at AT&T Bell Laboratories, that has been formulated around a centralized database. The database is organized in terms of the relational model, which is to say that design data is represented as a collection of homogeneous tables. Among the collection of data managed by the VDD System is information about the implementation technology itself. The recognition of the importance of storing technology information within the database is one of its important contributions.

VDD has not been built on a general purpose relational database system, however. The relational or tabular data model has been implemented via an efficient access method based on B+-tree data structures over character files. The full generality of the relational model and its high-level data manipulation operations (such as the relational algebra) is not supported. The VDD database component has been interfaced to an ensemble of design tools, inluding browsers, a layout editor, a design rule checker, and a switch-level logic simulator.

For the purposes of this chapter, we are most interested in the structuring primitives of its design database. A design consists of interconnected *transistors,* which are four terminal devices: source, drain, gate, and substrate. Each terminal, or connection point, is found on a particular technology dependent layer, such as diffusion or polysilicon. A *wire* is a piece of single layer material that connects two points within the design. If the two points are on different layers, then two wires on the appropriate layers must be connected with a *contact cut* to create the interconnection. Islands of electrically connected materials of the same type are called *nodes.* In connecting two points on different layers, the interconnection passes through three nodes: the wire from the source to the contact, the contact, and the wire from the contact to the destination. The listing of interconnections in terms of wires between nodes is called a *netlist.*[6]

Design data is aggregated into *cells* of information. A cell is constructed hierarchically from other cells and primitive components. In the case of a VLSI specific database, the latter includes such objects as transistors, wires, and contact cuts. The list of points on the inside of a cell, used in describing the interconnection of primitives and subcells, is called the cell's *interns*. In addition, each cell has an associated interface: a list of input/output ports. This is called its *externs*. They provide the names of connection points that are used when incorporating an instance of the cell as a child within a higher level parent cell.

The hierarchical design structure is constructed by *calling* a child cell from within the definition of its parent. The call is parameterized with geometric transformation information describing how the child cell should be placed and rotated within the parent cell. Further, if the caller refers to the nodes of the callee – for example, to interconnect the cell instance with other cells or primitives– then it must specify a mapping between its internal node numbers and the external node numbers of the called cell.

An *element* of a cell is one of its components. It can be any of the following objects: a wire, an extern, an intern, a contact cut, a transistor, or a call to another cell. Each of these are given unique identifiers, and can be accessed directly.

The collection of VDD tables describing the design are the following:
(1) *cell* (cellname, #externs, #interns, flag, timestamp, bbox);
   Each record of the cell table describes a cell within the design. With each cell is a

---

[6] This notion of netlist is slightly different than that discussed in the previous section.

description of its name, external nodes, internal nodes, status flag, timestamp of last design rule check, and a bounding box.

(2) *transistor* (cellname, trid, trname, trtype, poly, diff, dnnet, dsnet, gnet, snet, trbox, place, ref, rot);
Each record of the transistor table describes a transistor within a cell of the design. The information associated with the transistor is: the cell that contains it; the type of transistor it is; its gate (polysilicon) area; its source and drain (diffusion) area; the node numbers of its drain, source, gate, and substrate terminals; the bounding box of the transistor; information describing where the transistor has been placed within the containing cell; and reflection and rotation information.

(3) *cut* (cellname, cutid, cuttype, cutbox, cutnet);
The cut table describes each contact cut contained within each cell of the design. The descriptive information includes the type of the cut, its size and location, and its node number within the cell.

(4) *extern* (cellname, extid, extname, signame, extloc, layer, extnet);
The extern table describes the input/output ports associated with each cell. An input/output port has a name and unique id, is associated with an external signal, is located somewhere on the periphery of the cell's bounding box, is available on a specific layer of the technology (e. g., poly, diff, or metal), and has an external node number.

(5) *intern* (cellname, intid, intname, signame, intloc, layer, intnet);
The intern table lists information about the internal nodes of all cells. It includes information on the containing cell, intern id and name, the associated signal name, the location of the node within the cell, its layer, and its net number.

(6) *wire* (cellname, wid, layer, box, wnet);
The wire table describes the wires of the design. It contains one line per wire, representing the containing cell, the wire id, its layer, its size and location, and its net node number.

(7) *call* (callid, who, whom, instancename, place, ref, rot);
The call table describes what cells are parts of other cells. For each instance of a call to a cell, a line is placed in the table describing the calling cell, the called cell, and how the called cell is to be placed, reflected, and rotated within the caller.

(8) *callnet* (callid, instancename, oldnet, newnet);
The callnet table desdribes the remapping of node numbers whenever a cell is placed within the context of another cell.

The *call table* describes the hierarchical structure of the design, while the *transistor, cut,* and *wire tables* describe the design's representational details. In addition to this information, VDD stores information about the implementation technology in a separate *technology database*. The information stored here includes; a metatable, describing the structure of the technology database itself; a global table, describing global parameters of the technology, e. g., the circuit simulation parameters; a layer definition table, defining the primitive layers of the technology, e. g., poly, diffusion, etc.; a design rule table, specifying the acceptable width, spacing, and overlap tolerances for the layers; contact definition tables, describing the allowable contact types; and transistor definition tables, specifying the allowable transistor types. Note that many tools, such as Design Rule Checkers, are being made table driven to make them more technology independent. Such parameterization tables must be made part of the database description.

3.2.2 A Design Data Manager (SQUID)

[KELL82] describes the database component of a sticks-based symbolic design system under development at the University of California, Berkeley, by Richard Newton and his students. The system includes a terminal independent viewport manager, a collection of tools for stick-based design, and a general purpose design manager called SQUID.

The key element of the SQUID database is the *cell*. These are logical aggregates of design data, and are represented by a collection of representation-dependent *views* of the cell. Besides support for net/instance list and geometric views (consisting of rectangle, line segment, polygon, label, and doughnut slice objects on specific layers), the system also supports a "stranger" view for those data representations that cannot be directly interpreted by the data manager. Since cells are composed of subcells, the hierarchical structure of the design can be constructed in this way. The system also supports structuring primitives for design *regularity,* in the form of cell arrays. A cell can be defined as an array of repeated subcells, perhaps with controlled overlap.

The design data structure is mapped directly onto a conventional hierarchical file system. Each cell corresponds to a directory within the file system, and each view is implemented as a file within the directory. The internal structure of these files have been highly optimized for the representation that they contain. For example, the netlist file is implemented by a complex linked internal structure that mirrors the interconnection of circuit nodes.

The SQUID designers claim that it provides a general framework for design tool data management. Fast access is provided to the set of views known to the system, since these are implemented with special performance oriented file structures. However, the system remains open because of its support for "stranger" views, allowing it to manage representations beyond those designed into the system.

3.2.3 Complex Objects (System-R)

System-R, an experimental relational database system developed at the IBM Research Laboratory at San Jose, is currently being extended to support engineering design applications. It corrects a number of known shortcomings of the relational model for representing complex structured data such as is found in design applications. As already mentioned in the previous chapter, relational systems have been designed for the management of regular formatted data. It is not easy to represent unstructured data, such as text or graphical images, in the relational model. Further, relational systems do not provide operations for manipulating hierarchically structured data. It is not possible to specify hierarchical relationships within the "pure" relational data model. Thus, it is difficult to insure that such relationships are well supported by the implementation of the database on disk.

36   Chapter 3. Design Data Structure

The solutions provided within System-R are facilities for (1) creating and manipulating *complex objects* and (2) supporting very long data items, i. e., those much longer than a physical disk page. We discuss complex objects first. The usual collection of relational domain types, e. g., integers, strings, etc., has been extended to include the special types *identifier, comp-of,* and *ref*. They allow the designer of the database organization to make known to the system the hierarchical relationshipsamong tuples within relations.

Consider the simple Module-Parts database of Fig. 3.9. Each relation represents a design entity type: modules, parts, part functions, pins, and signals. A tuple within a relation represents an instance of an entity of that type, and is uniquely identified by its *identifier* attribute (e. g., MODULES(MID), PARTS(PID), FUNCTIONS(FID), PINS(PIN), SIGNAL(SID)). To represent, for example, the hierarchical relationship that a part is a component of a module, the part tuple must include the identifier for its parent module entity (this is sometimes called *referential integrity* in the database literature). Attributes that contain identifiers of other relations are of type *comp-of* (also known as *foreign keys*). Examples include PARTS(MID), FUNCTIONS(PID), and PINS(FID). The comp-of attributes can be used to partition the relations into collections of entities that share the same parent in the hierarchy (see Fig. 3.10), and this could form the basis of the clustering of the tuples on disk. Since a given tuple cannot be clustered on more than one parent, one of the parent-child relationships will be "stronger" than the others. The weaker relationships are represented by *ref* attributes. For example, the tuples of PINS are more strongly grouped by their relationship with FUNCTIONS, but there is still a parent-child relationship with SIGNALS. This is represented by PINS(SID).

In addition, System-R has been extended to better handle unstructured data. The system includes a facility for creating and manipulating special "large" data items (i. e., greater than 2 gigabytes in length). Essentially, a general purpose byte-oriented file system capability has been embedded within the system.

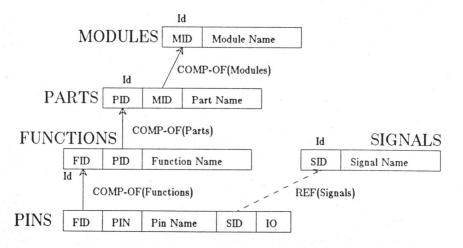

Fig. 3.9 Module-Parts Database Schema in System-R. The sample database consists of Modules, composed of Parts, which in turn are composed of Functions. Pins interconnect Functions and Signals

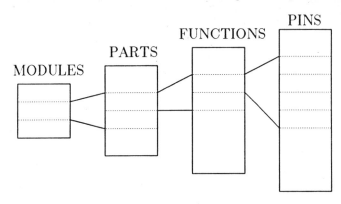

Fig. 3.10 COMP-OF Partitioning of Relations. All component tuples with the same parent tuple in the containing table are placed within the same partition. For example, part tuples used in the same module are grouped together in the Parts relation

The system is currently being used as the database component of a cell-oriented VLSI design system being developed at the University of Utah [HOLL84]. The system ties together the geometric layout and logic representation of a design. A design is recursively constructed from other designs and primitive cells. An interesting feature of the database is that it also stores the simulation stimuli and results, as well as the behavioral representation of the design in the form of an executable program. The relational schema is as follows:

(1) *Design* (Design-ID(ID), Design-Name, Version, Timestamp, Technology, Symbol(LONG));
Each entry in this table identifies a unique design object. Note that the graphical representation of the object, suitable for display on a graphics screen, is stored in the long data field Symbol. Tuples in this table are parents of tuples in the following tables: *Design-Blocks, Cell-Blocks, Connections, Design-Documentation, Design-Simulation, Design-EC-Log*.

(2) *Design-Usage* (Ident(REF), Use-Count);
Ident refers to a design object through Design-ID. Use-Count contains the number of other designs that incorporate the object. This is usaged primarily for the purposes of integrity maintenance. It is stored in a separate table to reduce contention for access to the design table.

(3) *Design-Blocks* (Design-ID(COMP-OF(Design)), Lower-Design(REF), X-Placement, Y-Placement, Rotation, Mirror);
Each entry represents the instance of a design object incorporated within another design object. The tuple encodes the object's geometric placement and orientation within the parent design.

(4) *Connections* (Design-ID(COMP-OF(Design)), Origin, Destination, EL-Properties(LONG), Constraints(LONG));
This table describes interconnection information associated with blocks, such as (1) the placement of connection pins around the periphery of a block; (2) electrical properties of the interconnection media; and (3) relevant constraints for use by the routing software.

(5) *Design-Documentation* (Design-ID(COMP-OF(Design))), Description-(LONG));
The table stores the documentation text associated with a design in the long field Description.

(6) *Design-Simulation* (Design-Sim-ID(ID)), Design-ID(COMP-OF(Design)), Simulation-Code(LONG));
The table stores the behavioral specification of a design object in the form of a program that simulates the object's behavior. associated design object.

(7) *Design-Simulation-Results* (Design-Sim-ID(ID)), Simulation-Method, Timestamp, Input-Sequence(LONG), Results(LONG);
The table stores both the simulation input vectors and the resulting output vectors in long fields.

(8) *Design-EC-LOG* (Design-D(COMP-OF(Design)), EC-Number, Engineer, Approval, Problem(LONG), FIX(LONG);
The table stores records of all changes made to a design object over time. Note that the description of the problem and its correction are stored in long fields within the table.

(9) *Logical-Cell* (L-Cell-ID(ID), L-Cell-Name, Cell-Set, Physical-Cell, Orientation, Graphics(LONG);
The Cell-Blocks table (described below) refers to Logical-Cell to incorporate cells within a design. Designers refer to logical cells which in turn are

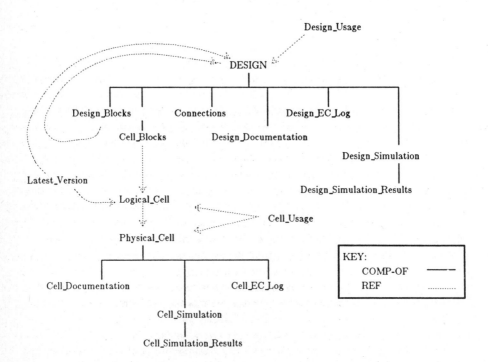

Fig. 3.11 Cellular Database Schema. A graphical representation of the Cellular Database Schema. Heavy lines represent partitioning relationships. Dotted lines represent relationships that do not partition, and thus, do not cause tuples to participate in a complex object

implemented by physical cells (see below). A given physical cell may have more than one logical cell counterpart. The table specifies how the physical cell is oriented to form the logical cell. The Graphics field stores a displayable symbol for the cell.

(10) *Physical-Cell* (P-Cell-ID(ID), P-Cell-Name, Version, Designer, Timestamp, Technology, Layout(LONG), Trans-Design(LONG);

The actual cell descriptions are found in this table. The cell's layout and schematic descriptions are stored in long fields.

(11) *Cell-Blocks* (Design-ID(COMP-OF(Design)), Cell(REF(Logical-Cell)), X-Placement, Y-Placement,Rotation, Mirror);

Similar to Design-Blocks, this table describes the placement of primitive cells within designs.

(12) *Latest-Version* (Name, Type, Cell-Ptr(REF(Logical-Cell)), Design-Ptr (REF(Design));

This table maintains pointers to the current versions of all logical cells and design objects.

Note how the structural information, namely the hierarchical construction of the design and the interrelationships across the layout and logical representations, have been implemented by the complex object mechanism of System-R. The representational information has been handled primarily by storing it as uninterpreted long data items. The design schema is represented graphically a shown in Fig. 3.11. It contains two independent complex objects: design and physical cells. Note how these form the roots of the COMP-OF hierarchy.

3.2.4 Abstract Data Types (Stonebraker)

INGRES, a relational database system developed at the University of California, Berkeley, has also been extended to support design applications, but in a different manner. INGRES supports an *abstract data type* (ADT) facility. ADTs are packages of "encapsulated" data structures and operations for their manipulation.They are encapsulated in the sense that the detailed implementation of the data structure is hidden from the user of the data structure. He can have access to it only through the operations provided by the abstract data type. While a relation can be implemented as an abstract data type, the concept is more useful for design applications if columns of relations are abstract data types. This is the approach taken in INGRES.

Consider a simple geometry database consisting of the following relations:

(1) *cell master* cell-id, cell-name, author, timestamp);

The table contains the cells of the design, their names, authors, and time of creation.

(2) *box* (cell-owner, layer, x1, y1, x2, y2);

The table contains rectangles of geometries, and records the cell that contains the rectangle, the layer of material it is on, and a specification of the rectangle in terms of its diagonal end-points.

(3) *wire* (cell-owner, layer, wire-id, width, x1, y1, x2, y2);

This table contains wire geometries. There will be one entry in the table for each segment of each wire. All segments of the same wire are related through a common value in the wire-id column.

(4) *polygon* (cell-owner, layer, polygon-id, vertex#, x, y);

The table is similar to the wire table, except that it stores vertices of layout polygons on a row by row basis. Vertices of the same polygon are identified by a common value in the polygon-id field.

(5) *cell-ref* (parent, child, instance-id, t11, t12, t21, t22, t31, t32);

This table represents the cell hierarchy. A child cell is represented as an instance within a parent cell. T11, t12, etc. specify a two dimensional transformation, rotation, and mirroring vector.

INGRES has been extended to allow the database designer to specify new column types and operations on these columns. For the sample geometry database, it may be useful to define the new column types box-ADT, polygon-ADT, wire-ADT, and array-of-floats. The database specification then becomes:

(1) *cell master* (cell-id, cell-name, author, timestamp);
(2) *box* (cell-owner, layer, box-desc);
(3) *wire* (cel-owner, layer, wire-desc);
(4) *polygon* (cell-owner, layer, poly-desc);
(5) *cell-ref* (parent, child, instance-id, orientation);

Tools built on top of the extended INGRES can manipulate the segments of a wire or the vertices of a polygon directly, rather than processing these a tuple at a time. New operations are also possible, for example, to compute the area of a box, or to determine if a point is contained within a box or polygon. The system supports user defined types and operators, thus types convenient to the tools being built can be defined and supported by the database system.

3.2.5 Semantic Data Model (McLeod's Event Model)

Semantic data models have arisen in an attempt to add more meaning to a database's data structure, by providing a richer set of modelling primitives than the tables of the relational model. [MCLE83] describes how an existing database semantic model has been extended and used to describe al VLSI design database. The model's primitives include *objects* and *application's events*. Objects are typed, and are related to other objects through *attributes*. Application's events are packages of manipulation commands that allow the database to be modified while maintaining semantic consistency.

To enable the model to better describe the data of design applications, it has been extended with primitives to support alternative implementations and evolutionary versions of design objects. The model has been augmented with:

(1) *AND nodes,* used to denote that an object "consists-of" an aggregation of successor objects. This construct can be used to describe composed-of and is-a-component-of relationship of the cell hierarchy.

(2) *OR nodes,* used to denoted that an object "is-implemented-by" one of the collected successor objects. This construct can be used to describe alternative object aggregations that represent different implementations of a design object.

(3) *LEAF nodes,* which cannot be further decomposed. This represents the leaf cells of the design hierarchy.

A sample database consisting of layout and circuit schematic data can be represented in terms of the following descriptor (i. e., character string) object types: event names, time stamps, instantiation ids, orientations, positions, user ids, version ids; and abstract (i. e., aggregation) object types: masks, wires, connections, netlists. A netlist object relates a net-id with a node name. A wire object relates a wire-id with a list of position pairs. A mask object relates a mask-id with a width, position, and layer description. A connection object relates a connection-id with a connection type and position. A version object relates a version-id to a collection of interrelated masks, wires, connections, and netlists.

It is straightforward to map the event model data description into a relational database data description. The set of relations and their columns are the following:

(1) *cell* (cell-name, tstamp, creator, default version, text);
Each line in this table represents the definition of a cell, who its creator is, when it was created, what its name is.
(2) *instantiations* (inst-id, orientation, child-cell-name, x, y, version-id);
This describes how instances of cells are incorporated within a particular version of a parent cell. The parent cell is identified by an entry in the version table.
(3) *versions* (version-id, cell-name, power, text, author, status);
Each row of the table represents a version of the implementation of a cell whose name is cell-name. The version-id field is used to tie together schematic primitives, netlists, wires, mask primitives, and instances of subordinate cells.
(4) *schematic-primitive* (version-id, prim-type, prim-id, inst-id);
Each row in the table represents a schematic primitive, such as logic gate, associated with a particular cell version.
(5) *netlists* (version-id, node-name, connection-id, connection-version);
A node is associated with a particular version of a cell through the version-it field. The node is connected to other nodes through connections, identified by their connection-ids.
(6) *connections* (connection-id, connection-type, position, version-id);
Connections are associated with specific cell versions through the version-id. They are associated with nodes through their connection-ids.
(7) *mask primitive* (mask-type, layer, wire-id, width-x, width-y, version-id);
Each row of the table represents a single mask primitive, its associated type, layer, and area. Mask primitives are associated with wires. A mask primitive is also associated with a given version of the cell in which it is contained.

3.3 Summary

Design databases are made complicated by their need to organize the many representations of the design. In this chapter, we have examined one design domain in detail, namely VLSI design, to show why there are so many different kinds of design descriptions. We have examined a number of different proposed and implemented design database systems, each taking a different approach for providing structuring primitives for the design database. System-R and INGRES are conventional relational database systems that have been extended in different ways to provide better support for engineering applications: System-R with complex objects and INGRES with abstract data types. VDD and SQUID provide solutions that are specific to VLSI design, and although database-oriented, cannot be considered to be true database systems. VDD organizes all its data as tables, while SQUID supports special purpose file structures for specific VLSI design representations. McLeod's Entity-Event model is yet another approach for describing design data as interrelated data entities. Yet, there is some commonality in the approaches, which we focus on in the next chapter.

4 The Object Model

4.1 Introduction

In this chapter, we give a detailed description of a semantic data model for design data called the *Object Model*. It defines the basic primitives from which the design data structure is formed. As we have seen in the previous chapter, many different schemes are possible, but similar themes are readily apparent. These include: (1) an applications interface based on the manipulation of „objects", which are structured collections of data, rather than individual records; (2) explicit support for design alternatives and versions, with operations to access the design description at a particular point in time; and (3) the critical importance of interface descriptions as a part of the design description. The objective of the Object Model is to make the design as selfdescribing as possible, thus enabling the design management system to assist in keeping its description consistent. It is interesting to note that a new draft standard interchange language, called EDIF (Electronic Design Interchange Format), is just beginning to emerge from the CAD system community, and bears a strong resemblance to the model presented in this chapter.

4.2 What are Design Objects?

Design objects are convenient aggregations of design information. They fall into two broad categories: *representation objects* and *index objects*. Representation objects describe a portion of the design in one of its representations, and thus are of a particular type, e. g., a layout object, a transistor object, or a logical object, as described above. Most information about the design is organized through representation objects. Index objects introduce auxiliary structures for grouping together representation objects and other index objects into useful clusters, e. g., objects that provide successive versions of the same "generic" object, such as an ALU.

Each representation object is constructed from the composition of its type's representation primitives (e. g., geometries, transistors, gates) and other objects of the same type. The hierarchical collection of design data thus formed is called a *representation hierarchy*. A *design hierarchy* is the collection of these describing a full-design, with additional structures linking together the representations and providing alternative groupings based on versions (configurations), alternative implementations, or simply attributes in common across the objects. Besides composition information, representation objects have *interface descriptions,* describing their abstract behavior, usage information, and associated performance (speed, power, area). A "design update" creates a new version of the updated object. It does not overwrite an existing representation object, but creates a new object which represents the new version. Such "version objects" are identified by their name extended with a unique version number.

Index objects provide a way to group objects together beyond the normal hierarchical decompositions within representations. Index objects are not limited to grouping together objects of the same type; they can also group together heterogeneous collections of objects. For example, a browsing application could make use of index objects to find objects with similar attributes, e. g., all ALU objects within a library could be grouped together by a single index object. Indices can be composed in much the same way as representation objects, and can be used to construct object taxonomies. For example, to represent that a specific ALU **IS-A** ALU object **IS-A** Datapath object, we could include the specific ALU within an index for all ALUs, which is in turn included within an index for all datapath objects. In the current specification of the data model, index objects do not have interfaces or versions. In the future, we may allow component object to inherit certain attributes from the indices that contain them. A number of distinguished index object types, providing additional structure to the design, will be introduced below.

The hierarchies can be represented as directed acyclic graphs (DAGs). Vertices represent objects. Leaves are *primitive* objects, while internal vertices are *composite* objects. The determination of what constitutes a primitive object from the viewpoint of the Object Model is left to the design. While these could correspond to individual representation primitives, e. g., a geometry or a transistor, a primitive Object Model object is more likely to correspond to some simple function implemented with a small collection of these. A composite object is formed from the recursive composition of its descendents in the graph. Edges in the graph are directed from *composite* (parent) objects to *component* (child) objects.

Associated with each edge is a specification of how to create an instance of the component object. Sometimes objects must be fully instantiated, creating a tree rather than a DAG. For example, when simulating a design, each instance is unique because each has different associated state variables. Only information that is unique to an object's instance needs to be represented in the tree (see Fig. 4.1).

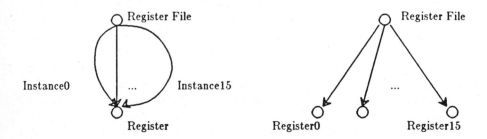

Fig. 4.1 Uninstantiated vs. Instantiated Objects. Instantiation information is associated with edges in the object graph. A design is instantiated by using this information to map the DAG representation into a tree. The single REGISTER object in the uninstantiated form is mapped into sixteen individual instances in the instantiated form

Although an object is defined by the composition of its components, it can be modified independently of them, and vice versa. A new version of the composite object can be formed from new components. The creation of new versions of the components does not affect their original composite. It can continue to reference their original versions. If the composite object's designer wishes to take advantage of the new versions, he must create a new version of the composite that explicitly includes them (see Fig. 4.2).

Sometimes design objects are only of interest as components of other objects, and do not stand on their own. To model this, objects have the attribute of being either *independent* or *dependent*. Independent objects exist within the database whether or not they are contained within other objects. Dependent objects are deleted from the database whenever they are no longer referenced. For example, a register object dependent on an independent register file object is of interest only as long as the register file object remains in the database.

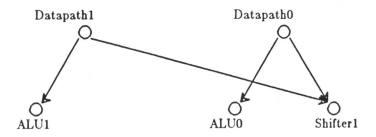

Fig. 4.2 Old Shifter, New ALU within a New Datapath. Objects carry their own configuration information. The original Datapath is composed from the original ALU and Shifter objects. A new Datapath incorporates a new ALU, but shares the original shifter

Representation hierarchies need not have identical decompositions. The functional decomposition of a design may be quite different from its physical decomposition. For example, in gate array technology, a given function could be implemented by a randomly placed collection of interconnected logic gates. The key observation is that there is no correlation between the placement of gates and the function they perform, thus no structure can be shared among these. Additional structure is needed to identify equivalent objects across representations is furnished by a special index object type, an *equivalency* object. These tie together objects in different representation hierarchies, constraining them to be equivalent. For example, an ALU layout object and an ALU transistor object may be linked by an equivalency object to denote that they are different representations of the same ALU. The constraints are enforced by the design validation component of the design management system (to be presented in Chap. 6). Equivalency objects are removed from the database when one of their referenced objects is removed.

Generic objects are another special type of index object. Along with index objects, they are "gateways" to the design (see Fig. 4.4). They represent major subsystems undergoing frequent change and refinement. They are independent objects, and can exist whether or not they are referenced by other objects. Additional structures for organizing alternatives and versions are used in their definition. A *version* configures

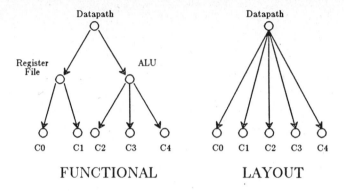

Fig. 4.3 Non-Isomorphic Representations. The object hierarchical decompositions cannot be constrained to be isomorphic across all representations. For example, the functional decomposition of a system may bare no resemblance to the decomposition used in its physical realization

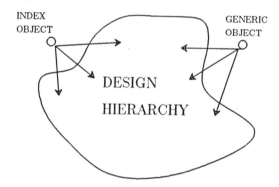

Fig. 4.4 Gateways to the Design. Index objects and generic objects are the entry points into the complex interconnected web of objects that form the design description

the subsystem, by correlating versions of representation objects describing the same subsystem, but in its different representations. Version objects are a special type of equivalency object, since the objects it groups together are constrained to be equivalent (i. e., they all implement the same subsystem). They differ in that versions exist independently of the objects they correlate. An *alternative* object is an index that groups together version objects, representing different versions of the alternative. Finally, a generic object (see Fig. 4.5) groups its alternatives together. Each alternative has the same behavior as its generic parent, but its own performance characteristics. For example, the generic object "the ALU" is composed of *alternative* objects "fast ALU", "small ALU", and "low power ALU". The "fast ALU" alternative consists of various version objects, e. g., "fast ALU/version 0.0", "fast ALU/version 1.0", etc. "Fast ALU/version 1.0" in turn consists of the representation objects describing it, e. g., "fast ALU/version 1.0/layout", "fast ALU/version 1.0/transistors", and "fast ALU/version 1.0/gates".

Chapter 4. The Object Model   47

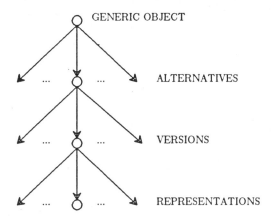

Fig. 4.5 Generic Object Data Structure. A *generic object* anchors the collection of objects that together describe a major subsystem of the design. It groups together *alternatives,* each of which anchor different implementation approaches for the same design object. These, in turn, group *versions,* which group together the evolutionary descriptions of a particular alternative implementation

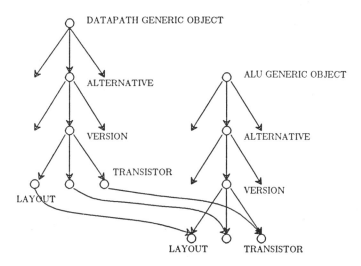

Fig. 4.6 Nesting of Generic Objects. The generic, alternative, and version objects are additional structure imposed on the interconnected collections of representation objects. They form hierarchies that are orthogonal to the representation hierarchies. Generic objects can only be nested at the lvel of representation objects

Version objects, alternative objects, and generic objects offer a mechanism by which convenient groupings of representation objects can be formed. This is in addition to the representation hierarchies. Generic objects are nested on a representation by representation basis (see Fig. 4.6). An *implementation* of a generic object is a version of one of its alternatives. To nest an ALU within a datapath object proceeds as follows. An ALU implementation is chosen for inclusion within a datapath implementation. Each of its grouped together representation objects, e. g., the layout, transistor, and gate objects, incorporate the object of corresponding type in the ALU's implementation.

48   Chapter 4. The Object Model

Designs are very special generic objects. A design is the root of a hierarchical collection of design objects, the creation of which is the objective of the design team. For better control of the design data, arbitrary sharing of objects across designs is not allowed. *Library objects* provied the mechanism through which design objects can be shared. Only library objects can be shared among designs (see Fig. 4.7). These are generic objects of general utility, designed specifically for incorporation within other objects. The decision to place an object into a design library should be made carefully. For example, the input/output pads used in almost every design are library objects.

To make the design specification more concrete, we will describe the simple shift register introduced in Sect. 3.1 in terms of the object model. Assume that the shift register is composed of two instances of a half shift register, each consisting of a clocked pass transistor and an inverter. The components of the description include the object's name, its version and designer, its type, the time of its creation, the names of the other objects that incorporate it as a component, its interface specification, how it is composed from more primitive objects, and finally, its representation details.

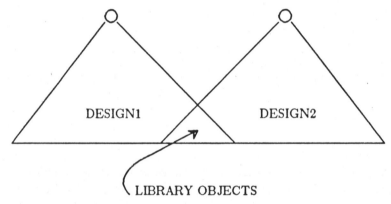

Fig. 4.7   Designs and Libraries. Only library objects can be shared across designs. Care must be taken to ensure that only objects of general utility will be placed in libraries

We have adopted an *s-expression* or parenthesized notation for writing down the object specification. These are built from nested expressions of the form (*SLOT-NAME value1 value2 value3* . . .), where each value can be a simple atom or a parenthesized expression. S-expressions are used because the format is easy to extend with new slots. Also, because the syntax is similar to the LISP programming language, a design specification constructed in this way could eventually be executable as well as descriptive. The skeletal (sometimes called "frame") representation object specification for the shift register cell looks like the following:

```
(NAME ShiftRegisterCell)
(VERSION 1.0)
(DESIGNER Randy H. Katz)
(TYPE Independent Representation Layout)
(TIME Mon May 2 21:43:15 CDT 1983)
(WITHIN (RegisterCell 1.0) (RegisterCell 2.0))
(INTERFACE ... )
(COMPOSITION ... )
(REPRESENTATION ... )
```

The details of the INTERFACE and COMPOSITION portions of the specification will be given in the following subsections.

4.3 Interfaces: How to Use a Cell Without the Details

Interface descriptions are associated with each representation object. An interface should contain enough information about the object so it can be used without having a detailed understanding of its implementation. It documents the function of the object and makes connectivity information explicit. Interface descriptions include the following:

(1) **Name:** the name of the object, including its version.

   (NAME ShiftRegisterCell)
   (VERSION 1.0)

(2) **Designer:** the designer who is responsible for its implementation (not neccessarily the designer who implements it).

   (DESIGNER Randy H. Katz)

(3) **Description:** a description of the object's behavior (e. g., an English-language description, a truth table describing outputs in terms of input combinations, etc.). This description is primarily for documentation purposes, but could eventually be used by validation tools as the appropriate specification for simulation.

   (DESCRIPTION 2 Phase Dynamic Shift Register Cell)

(4) **Graphical Representation:** Objects are contained within *bounding polygons*. This is a useful way to specify the "outline" of the object for viewing on a graphics terminal. Bounding polygons can be associated with any representation type.

   (POLYGON (0 0) (0 20) (20 20) (20 0))

(5) **Connectivity Information:** *Ports* are the input and output connection points of an object. Ports are named, have types and directions, and are placed on the periphery of the bounding polygon. The same name can be assigned to more than one port. Ports so named must be connected by some signal path within the object. The directionality of the ports are input, output, and bidirectional. The type information depends on the representation. For example, for the layout representation, type information includes interconnection layer and the form of logic level expected or supplied by the port.

```
(PORTS
    (LOCAL PORTNAME In DIRECTION Input TYPE 4:1 LOCATION (0 10))
    (LOCAL PORTNAME Out DIRECTION Output TYPE Gate LOCATION (10 10))
    (GLOBAL PORTNAME Phi1 DIRECTION Input TYPE 4:1 LOCATION (5 20))
    (GLOBAL PORTNAME Phi2 DIRECTION Input TYPE 4:1 LOCATION (15 20))
```

(6) **Performance Information:** Depending on the representation, the interface specifies constraints on power, area, and delay exhibited by the object in performing its function.

```
(PERFORMANCE (DELAY 5 ns) (AREA 42 um BY 42 um) (POWER 10 uw))
```

The interface contains information that is used to verify that a design is self-consistent. Type systems simplify checking that object compositions are well-formed. One particular type system has been used in the Stanford Cell Library [NEWK81]. The input types are: (1) 4:1 ratio (expects regenerated logic levels); (2) 8:1 ratio (can tolerated degraded logic levels); (3) switch control (expects regenerated logic levels); and (4) switched (can tolerate degraded logic levels). Output types are: (1) gate (produces regenerated levels); (2) superbuffer (produces regenerated levels); (3) switch logic (produces degraded levels), and (4) precharged (special). The compatibilities among the types is determined from a table, such as Table 4.1. Other type system are also conceivable.

Table 4.1 Compatibility of I/O Types

Inputs	Outputs			
	Gate	Superbuffer	Switch Logic	Precharged
4:1	O.K.	O.K.	NO	O.K.
8:1	O.K.	O.K.	O.K.	O.K.
Switch Control	O.K.	O.K.	NO	O.K.
Switched	O.K.	O.K.	(1)	NO

Note (1): O. K., but charge sharing problems are possible.

Since design objects can be created independently of other portions of the design, it is important to provide a scoping mechanism for the naming of ports. We have adopted a simple approach. Port names can be either *local* or *global*. The ports of two objects with the same local port names are different, while every port with the same global name is implicitly connected. Vdd and GND are typical global port names.

The interface constrains the object's implementation. An important task of design validation is to insure that the object's implementation meets its design specification. Checking that power and area are within constraints is straightforward: they are associated with the layout representation and can be computed from the cell's bounding box of geometries and the width-to-length ratios of its transistors. Delay (timing) is more difficult to specify, and, in general, can only be checked through detailed timing simulations.

It is often desirable for the designer to be able to specify an object's interface before its actual implementation. In conjunction with the appropriate design tools, such as a multi-level simulator, the designer can test a portion of the design in context by given rough specifications for its performance. The performance information is approximate, and is specified within ranges. In this case, the interface specification is a "promise" about how the object will eventually perform. Once an object has been implemented, its performance can be readily determined.

Describing the behavior of an object for documentation purposes is difficult, because of the many ways to describe behavior. Hardware description languages for providing complete and convenient ways to specify an object's behavior are still a research topic. For a given object, one form may be more appropriate than another. Truth tables or logical expressions are well-suited for describing combinational logic. Transition tables or state diagrams are appropriate for sequential logic. Alternatively, an object can be associated with a program that "simulates" its behavior. Many language-based functional simulators use this approach. An approach suitable for functional/timing simulations, or for circuit testing, describes the behavior of a module by the input waveforms and expected output waveforms. Maintaining detailed waveforms may be expensive in storage space, but is the easiest way to specify the expected timing behavior of a module. The description portion of the interface must be general enough to support any of these. At the very least, it can be an uninterpreted character string. This is one of the strengths of choosing the S-expression notation for our design database description: it can be adapted to future behavioral specifications as they get formulated.

Showing that an object's implementation agrees with its interface includes the following verification tasks. Implied connectivity between ports must be verified, as must be their types and directionality. An "interface extractor" program could aid in the verification. Note that implementations do not need to be checked if these have been derived from their interfaces ("correctness by construction"). One could envision a PLA generator that takes the interface's behavioral description, perhaps specified as a set of Boolean equations, along with the locations of input, output, power, and clock ports, and generates the PLA layout within the object's bounding polygon.

The complete interface description looks like:

```
(INTERFACE
        (POLYGON (0 0) (0 20) (20 20) (20 0))
        (PORTS
                (LOCAL PORTNAME In DIRECTION Input
                        TYPE 4:1 LOCATION (0 10))
                (LOCAL PORTNAME Out DIRECTION Output
                        TYPE Gate LOCATION (10 10))
                (GLOBAL PORTNAME Phi1 DIRECTION Input
                        TYPE 4:1 LOCATION (5 20))
                (GLOBAL PORTNAME Phi2 DIRECTION Input
                        TYPE 4:1 LOCATION (15 20))
        )
        (DESCRIPTION 2 Phase Dynamic Shift Register Cell)
        (PERFORMANCE (DELAY 5 ns) (AREA 42 um BY 42 um)
                (POWER 10 uw))
)
```

52   Chapter 4. The Object Model

The interface description for the half register cell looks like:

    (INTERFACE
        (POLYGON (0 0) (0 20) (10 20) (10 0))
        (PORTS
            (LOCAL PORTNAME In DIRECTION Input
                TYPE 4:1 LOCATION (0 10))
            (LOCAL PORTNAME Out DIRECTION Output
                TYPE Gate LOCATION (10 10))
            (LOCAL PORTNAME Clk DIRECTION Input
                TYPE 4:1 LOCATION (5 20))
)
        (DESCRIPTION Half Of Dynamic Shift Register)
        (PERFORMANCE (DELAY 2 ns) (AREA 42 um BY 21 um)
            (POWER 5 um))
)

4.4 Composition and Interface

Composing objects to form composite objects can be viewed in graphical terms. Placed within the composite's bounding polygon are the polygons of its components. These must be "composed" in some fashion to implement the higher level function of the composite object. This is done by wiring the components together, either by *interconnection* or by *abutment*, and by identifying which parts of the components are mapped into the ports of the composite. The latter is indicated by wiring a component port to a composite port.

For example, consider Fig. 4.8. Instances X and Y of the HalfShiftRegister are composed to form a ShiftRegister object. The Out port of X is connected to the In port of Y. The other ports of X and Y are mapped onto the ports of the ShiftRegister. For example, In of X becomes In of the ShiftRegister while Out of Y becomes Out of the ShiftRegister.

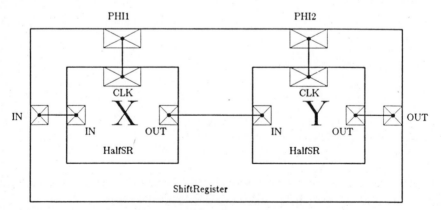

Fig. 4.8 Graphical Composition: Wiring By Interconnection. The bounding polygons for two instances of the HalfSR are incorporated into the ShiftRegister's polygon. The interconnection among ports is shown explicitly by wiring. The figure shows a graphical way of displaying the hierarchical relationship among parent and child cells. Its use is not limited to those representations that are normaly displayed graphically

Besides interconnection information, the object must include information about its components and where it is used as a component. for example, if ShiftRegisterCell is contained within versions 1.0 and 2.0 of the Register object, then its specification would include the following:

(WITHIN (Register 1.0) (Register 2.0))

A unique instance name is associated with each component, and the composite describes how each is oriented and placed within its bounding polygon. The S-expression description of the example of composition of figure 4.8 would appear as the following within the ShiftRegisterCell:

```
(COMPOSITION
   (INSTANCE x NAME HalfRegister TRANSLATED (0 0))
   (INSTANCE y NAME HalfRegister TRANSLATED (10 0))
   (INTERCONNECT
     ((y In) (x Out))
     ((x In) (ShiftRegisterCell In))
     ((y Out) (ShiftRegisterCell Out))
     ((x Clk) (ShiftRegisterCell Phi1))
     ((y Clk) (ShiftRegisterCell Phi2))
   )
)
```

Since the Half Register Cell is primitive, i. e., it has no component objects, its composition specification would be empty:

(COMPOSITION)

Similarly, if the Shift Register Cell contains no layout geometries, its representation specification would be left empty.

4.5 Complete Example of Object Specification

Here we collect together the complete description, except for representation details, of the ShiftRegisterCell and HalfShiftRegisterCell design objects:

```
(NAME ShiftRegisterCell)
(VERSION 1.0)
(DESIGNER Randy H. Katz)
(TYPE Independent Representation Layout)
```

```
        (TIME Mon May 2 21:43:15 CDT 1983)
        (WITHIN (Register 1.0) (Register 2.0))
        (INTERFACE
            (POLYGON (0 0) (0 20) (20 20) (20 0))
            (PORTS
                (LOCAL PORTNAME In DIRECTION Input TYPE 4:1 LOCATION (0 10))
                (LOCAL PORTNAME Out DIRECTION Output TYPE Gate LOCATION (10 10))
                (GLOBAL PORTNAME Phi1 DIRECTION Input TYPE 4:1 LOCATION (5 20))
                (GLOBAL PORTNAME Phi2 DIRECTION Input TYPE 4:1 LOCATION (15 20))
            )
            (DESCRIPTION 2 phase dynamic shift register cell)
            (PERFORMANCE (DELAY 5 ns) (AREA 42 um BY 42 um) (POWER 10 uw))
        )
        (COMPOSITION
            (INSTANCE x NAME HalfRegister TRANSLATED (0 0))
            (INSTANCE y NAME HalfRgeister TRANSLATED (10 0))
            (INTERCONNECT
                ((y In) (x Out))
                ((x In) (ShiftRegisterCell In))
                ((y Out) (ShiftRegisterCell Out))
                ((x Clk) (ShiftRegisterCell Phi1))
                ((y Clk) (ShiftRegisterCell Phi2))
            )
        )
        (REPRESENTATION)
)
(
        (NAME HalfRegister)
        (VERSION 1.0)
        (DESIGNER Randy H. Katz)
        (TYPE Dependent Representation Layout)
        (TIME Sun May 1 09:31:15 CDT 1983)
        (WITHIN (ShiftRegisterCell 1.0))
        (INTERFACE
            (POLYGON (0 0) (0 20) (10 20) (10 0))
            (PORTS
                (LOCAL PORTNAME In DIRECTION Input TYPE 4:1 LOCATION (0 10))
                (LOCAL PORTNAME Out DIRECTION Output TYPE Gate LOCATION (10 10))
                (LOCAL PORTNAME Clk DIRECTION Input TYPE 4:1 LOCATION (5 20))
            )
            (DESCRIPTION Half of a dynamic shift register)
            (PERFORMANCE (DELAY 2 ns) (AREA 42 um BY 21 um) (POWER 5 uw))
        )
        (COMPOSITION)
        (REPRESENTATION Boxes
            (Metal (0 0) (10 4))
            . . .
            (Poly (4 20) (6 0))
        )
)
```

4.6 Objects Implemented as Structured Files

Objects can be implemented as files extended with information describing the design data structure. The information relevant to the design management system is the object type, the interface specification, and the composition information (see Fig. 4.9). Representational details are determined by the design tools, not the design data management system. We expect that new tools will be created that combine the creation of the interface, composition, and representational specifications. Existing design files can be referenced from within design object files to include their data in the design data structure.

Representation
Object File
Format

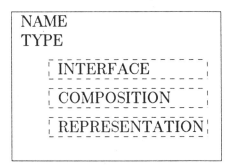

Fig. 4.9 Design Objects as Files. Design objects can be mapped into conventional files in a staiightforward way. The important point is that we are imposing a particular *structure* on these files that will be of use to the design management system

5 Design Transaction Management

5.1 Introduction

A *transaction* is a sequence of operations that together perform a unit of work, and leave the database in a consistent state. A set of constraints are in force at the beginning of the transaction and again at its end, but they are not enforced during the lifetime of the transaction. A key difference between the design environment and a conventional database environment is that the definition of design self-consistency is significantly more complex. For example, an airplane design is "consistent" only if the airplane can still fly with its redesigned wing. In general, these constraints can only be guaranteed by invoking complicated checking programs such as simulation tools.

The database community has been very successful in developing sophisticated implementation mechanisms to support transaction in "transaction processing environments", such as airlines reservations and electronic funds transfer. These applications are characterized by high volume, short duration, simple units of work. Yet, design applications are considerably different. For example, the way in which design objects are accessed differs from conventional databases because of the long duration of design interactions as well as the dispersed computing environment that supports design. Complex engineered objects are designed by teams, simultaneously working on different portions of the overall design. One of the goals of a design management system is to support the controlled sharing of design data, through mechanisms to insure that designers do not interfere with one another. Further, design data is a valuable resource; it must survive across even catastrophic system crashes.

The database component of the design system provides *a shared* repository for design data that is *resilient* to system crashes. The collection of mechanisms that provide controlled sharing and crash recovery are called *transaction management*. However, the conventional notions of transaction consistency, atomicity, and durability, developed for transaction processing, must be extended for the design environment. The issues in providing such mechanisms in the design environment are discussed in this chapter.

Our purpose in this chapter is to examine how design transactions differ from conventional transactions, and how that difference affects the transaction management component of a design management system. We begin with a description of the computing environment for design in Sect. 5.2, and then compare conventional and design transaction models in Sect. 5.3. We review the concurrency control (Sect. 5.4) and recovery (Sect. 5.5) issues in the design environment, and use this to motivate the definition of a simple design transaction model (Sect. 5.6). Some extensions to the model are presented in Sect. 5.7, and related work is reviewed in Sect. 5.8.

5.2 Design Computing Environment: Implications for Data Management

With the advent of inexpensive engineering workstations, the environment in which design activity takes place is changing rapidly. A large time-shared mainframe computer is no longer the standard configuration. It is being replaced by a high-speed network of engineering workstations connected to special purpose hardware servers. A typical workstation consists of a general purpose microprocessor, reasonably high quality graphics, and usually a local disk. The special purpose hardware includes dedicated file server machines, and perhaps special hardware for computationally intensive aspects of the design process, such as design rule verification and simulation.

The following scenario is typical of the design environment. Designs are created at workstations, connected by a high speed network to a shared database server. The design is created and manipulated at the workstations, with the server providing a centralized location for long term data storage. The workstations have simple I/O configurations, typically a single Winchester disk, while the server has a large number of disk devices on multiple channels. Designers at the workstations transfer the relevant parts of the design to their local disks, through a design transaction mechanism described below. Work proceeds independently at the stations, with incremental changes spooled to the server, providing back-up in case of a workstation crash. The changes are incorporated into the public database only after they have been shown to be valid.

The dispersed database is actually more complicated than what we have outlined above. Some validation can be performed at the workstation, but much is done on special purpose hardware or higher performance mainframes. Also, data may be shared among organizations that are geographically distributed, e. g., a design center in San Jose and a manufacturing facility in Phoenix. The database server must keep track of the location of design data at all times. Throughout this chapter, we concentrate on database server/workstation interaction, but the problems and solutions apply in the more general environment.

5.3 Conventional Transactions in the Design Environment

In this subsection we review the models of data access in the conventional transaction processing environment, and contrast this with the way in which design data is accessed. To maintain similarity with the database notion of transaction, we define a *design transaction* as a sequence of database operations that map a consistent version of a design into a new consistent version. Design transactions are non-atomic units of design consistency. If the system crashes, then the designer can continue from the last safe state determined by transaction management (which may be beyond the last saved state). Old design versions are durable across transactions, i. e., an old version is not removed unless it is explicitly moved off-line.

The canonical "debit/credit" transaction is typical of conventional transaction processing. The database consists of a collection of bank account, teller cash drawer, bank branch balance, and account history records. Tellers handle customer deposits and withdrawals in real time. A typical transaction, many of which are executing simultaneously, is invoked on behalf of a teller. It accesses a single account record, checks the balance to insure that there is sufficient funds if a withdrawal, modifies the balance, and makes the modified record available to other transactions. The teller cash drawer and branch balance records are also modified by the transaction. A history record is created to provide an audit trail of the transaction. Disastrous results ensue if more than one transaction is allowed to modify these records simultaneously, i. e., updates may be lost and the database may be left in an inconsistent state. A transaction processing system must guarantee high throughput and fast response, even when the size of the database is very large.

On the other hand, a typical "design trans(inter)action" behaves as follows. The transaction is invoked by a designer to extract a logical portion of the design from the shared design repository into his private workspace. During the lifetime of the transaction, he interacts with his data through the ensemble of design tools available to him: design editors, generators, and the battery of analysis programs used to check the design for correctness.

When his design activities are complete, i. e., the designer believes that his data is once again consistent, he returns it to the shared repository to "release" it to other designers. Before he is allowed to do this, however, the data must pass a battery of tests for self-consistency. The validation process is complex, time consuming, and specific to the object being designed. The design system is responsible for insuring that all checks are performed in the desired sequence.

Once the design data has been shown to be self-consistent, it can be replaced in the design repository. However, older versions of a design file, perhaps containing the description of a design object, are rarely discarded once a new version has been created. These old versions may be needed (1) for legal purposes, (2) because they still describe a supported object installed in the field, or (3) to provide insights into the design process itself. The new version is added to those that are available on-line in the repository. *Conversational transactions,* discussed in [GRAY78], superficially resemble our *design transactions* since both are non-atomic. However, there are a number of differences. *Conversational transactions* have conventional transactions as units of recovery ("nested transactions"), while *design transactions* support continuous saving of design changes and have the ability to recover past savepoints. With *conversational transactions,* the effects of a nested transaction can be undone only by a compensating transaction. Because *design transactions* support versions, it is possible to simply restore the modified files to their previous versions. The need for applications level consistency checking further distinguishes *design transactions* from *conversational transactions.*

Given these two contrasting transaction models, we can draw several conclusions:

(1) **Design Transactions are long duration**

Designers interact with their data for long periods of time, i. e., days or weeks, while *conventional transactions* are of short duration, i. e., minutes at most. Thus, mechanisms that arbitrate exclusive access to shared data by forcing transactions to wait when it is not available are unsuitable in the design environment. Suspended transactions would be forced to wait for intolerably long periods of time. Real time access is critical in most transaction processing environments, yet designers are content to try again later to get the needed data. Further, *conventional transactions* spend most of their time in the database access routines, because the applications logic is relatively simple. *Design transactions* spend most of their time in the associated "number crunching" applications programs (i. e., the design tools). Thus, they are not as closely coupled to the database system as *conventional transactions*.

Long duration transactions in the design environment do not have the problems of long-lived transactions described in [GRAY81]. For example, visible intermediate transaction states can be tolerated (i. e., "lower levels of consistency" are acceptable). While long running conventional transactions are usually aborted on system restart, long duration *design transactions* need not, and should not be aborted.

(2) **Design Transactions touch large volumes of data**

The units of access in *design transactions* are large collections of related records, usually spanning several files. *Conventional transactions* are simple, and touch very few records. While *design transactions* spend a relatively small amount of time in the database system, the large volumes of data involved prohibit invoking the database system for access to individual records. Therefore, design databases are best used as shared repositories from which data must be extracted when needed for intensive access.

(3) **Design Transactions demand more than serial consistency**

Correct execution of concurrent transactions has been defined in terms of serializability, i. e., the execution of concurrent transactions is consistent if their interleaved effect is the same as if they are run in some serial order (i. e., one at a time, one after the other). Serial consistency is unsuitable for determining whether design data is still "consistent" after update. Special validation programs must be invoked to verify the consistency of design data. Serializability theory describes when interleaved read/write accesses to shared data still obtain "correct" (i. e., serially consistent) results. Yet, interleaved updates are undesirable in the design environment, since it is meaningless for two designers to change the same portion of a design.

(4) **Design Transactions are not all or nothing**

Conventional transactions are atomic: either all updates made by a transaction become visible (it commits) or none are visible (it aborts). Intermediate states are invisible to concurrent transactions, even if the system should crash during a transaction. Recovery mechanisms insure that the database is restored to a transaction consistent state. In the design environment, as much work as possible should be recovered in a crash, even past a checkpointed state if possible. Intermediate states, as long as they are file system consistent, are acceptable.

(5) **Design Transactions are not ad hoc**

Since designers know in advance what portions of the design they will be working on, all needed resources can be acquired at the beginning of the transaction. Due to the interactive nature of design transactions, deadlock is intolerable, and must be avoided through preallocation. Deadlock detection mechanisms that abort in-progress transactions are undesirable, since valuable design work would then have to be undone.

Design transactions do not fit the conventional notions of consistency, atomicity, and durability upon which database transaction management has been built. Serial consistency is insufficient for determining the self-consistency of design data. The data itself, rather than the order in which it is accessed, determines the correctness of a transaction. Further, simultaneous access to the same design data is unlikely as well as undesirable. The differences are summarized in Table 5.1, and are elaborated below.

Design transactions are not atomic in the sense of *conventional transactions*. Visibility of intermediate states of design data may even be desirable. While only one designer is allowed to update the data, many can be reading ("browsing") it simultaneously. Others may want to check on an in-progress portion of a design. Since they are browsing the design, a lower level of consistency can be tolerated, i. e., records can change underneath them.

Designers demand that as much of their work as possible be saved in the event of a system failure. Savepoints guarantee that changes are saved, but it is desirable to be able to bring the database back to its *latest* possible state. While returning to a checkpointed state should be supported, system generated undo will be rare. Some systems, such as [LORI82], only support recovery to a checkpointed state.

Table 5.1   Conventional vs. Design Transactions

	Conventional Transactions	Design Transactions
Atomicity	Transaction is Atomic	Transaction is Non-atomic
Consistency	Correctness in terms of Serial Consistency	Correctness in terms of Application-Level Consistency
Durability	Committed Changes Survive System Crash	Committed Changes Survive Across Time (Versions)

The durability of design transactions is also different from conventional transactions. Old versions of design data persist even after newer consistent versions have been created. Support for versions is already needed in the design environment, and can be integrated with transaction management to simplify many aspects of concurrent access and recovery. Transaction management components of existing database systems do not support versions, making these somewhat unsuitable as a starting point for design transactions support.

5.4 Concurrency Control Issues

Design data is arranged so that logically related parts of the design can be accessed as a single object. As described in chapter 4, object can be nested within other objects, forming a hierarchy of design data. A designer can request access to the whole design or any of its subparts. Designer can work in parallel as long as they are in non-overlapping subtrees of the design hierarchy.

Multiple designers do not work on the same related pieces of the design at the same time. Changes made by one designer might conflict with those of another. Even if these changes do not overlap, merging the sets of changes together might not result as intended, since the changes may not fit together. Therefore, the appropriate unit of exclusive access is a design subhierarchy, representing a logical portion of the design, and identified by its root.

Conventional database locking is not appropriate for design data. If a design subtree is unavailable because it has been acquired by another designer, then the requesting designer should not be forced to wait. It could take a rather long time for the designer who currently holds the data to return it to the repository. Also, because locks need to be held for long periods of time, they need to survive system crashes. The solution is to introduce *persistent locks* that are stored in non-volatile storage, i. e., the database, to survive crashes. These locks can be implemented so a transaction does not suspend execution in the event that the lock is held by another transaction.

Note that the lock table is normally placed in memory for reasons of efficiency. Design transactions are less time critical than conventional transactions. Placing the lock table on disk to insure its durability is worthwhile in the design environment, even though it becomes more expensive to set/reset locks.

A more appropriate paradigm for acquiring design data is to view the design database as a library. Design subparts are checked out to designers, who return them when done. "Concurrency control" is therefore handled by a remote Librarian process that traverses and manipulates the hierarchical structure of design data on behalf of the designer at his workstation. It knows: (1) *what* parts of the design have already been checked out; (2) *who* has checked them out; and (3) *when* they are expected to be returned to the repository. This information is stored in the design database, and is thus durable across system crashes. A designer who must have access to data can determine who currently holds it, enabling him to negotiate for its early return.

When a designer is through with his data, it is checked back into the repository as a *new* version. The modified design data cannot become the current version ("released"), until it has passed the necessary self-consistency checks. It can be accessed by other designers who are willing to browse the possibly inconsistent data (it may change underneath them). A transaction cannot complete, and cannot make its new versions available, until the data is shown to be consistent. At commit the *in-progress* versions of all data checked out to the transactions become *current* and the previously current versions become the most recent *previous* versions.

The facilities provided by the Design Libarian more closely resemble a time-domain addressing scheme than a conventional lock manager. A version-based approach is well-suited to the design environment, since versions of design data must already be supported. The versionbased approach eliminates many of the difficulties of concurrent access in the design environment. Designers can browse the last consistent version of the design even while a new version is inprogress. If desired, the in-progress version can be browsed, although reads may not be repeatable. Even if they are, the data read made yield inconsistencies, since it is not yet guaranteed to be consistent. Browsers can continue to read the same version even when the in-progress version becomes current in the middle of browsing.

5.5 Recovery Issues

Because of the value of design data, resiliency to system crashes is an important requirement of the database component. Conventional transactions restore the database to a transaction consistent state in the event of a crash. Design transactions are not atomic in this sense. If the system crashes in the midst of a design transaction, it may be undesirable to undo any work done by the transaction. Obviously, file system actions, such as page writes, must be atomic, and it most be possible to reconstruct the file system to a consistent state (from its viewpoint) after a crash.

If the design transaction system supports savepoints, it can avoid the problems of restoring the database to a transaction consistent state, by supporting recovery to savepoint consistent states. Note that this may not be enough, since it is still desirable to be able to recover past savepoints.

In addition to the demands for a continuous recovery capability, the computing environment for design introduces additional implementation problems. As described above, the design environment consists of individual workstations networked to a file server. They are inherently less resilient to crashes than the file servers, because they are located in a more hostile environment (an office instead of a machine room), and because it is more difficult to use redundancy to obtain resiliency (workstations rarely have more than one disk, and almost never have tape drives). The result is that the redundancy for data on the workstation must be provided by the file server.

The recovery system is made even more complicated by the increased number of ways in which the system can fail in a dispersed data environment. These include: (1) workstation soft crash (memory buffers lost); (2) workstation hard crash (local disk data lost); (3) server soft crash; and (4) server hard crash.

On the other hand, the nature of the design environment simplifies many recovery aspects. While design versions are checked out to workstations, the last consistent version resides safely on the file server. Versions are used to avoid updating in place, with its associated undo complexities. Only very limited undo capabilities are needed – for example, undo the last update. Transaction logs, or change files, have a simple structure and are associated with files rather than transactions, since only one transaction can update a design file at a time.

5.6 Design Transaction Model

Design transactions are the mechanism by which designers create new consistent versions of design objects. Design transactions consist of *work, validation,* and *completion* phases. While work and validation can be intermixed, validation must be complete before a design transaction is allowed to enter completion (see Fig. 5.1).

During the *work* phase, a designer requests design objects from the Design Librarian. If the object has not been granted to another designer, the request is honored and the appropriate files are transferred to the workstations's disk. Additional *mirrored* copies are made in the database server, providing redundant copies used for recovery purposes.

Chapter 5. Design Transaction Management

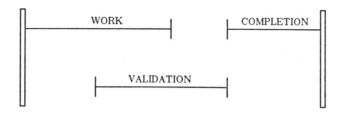

Fig. 5.1 Design Transaction Phases. A design transaction pass through a work phase to a validation phase to a final completion phase, when its changed objects become public. Validation must complete before completion can begin

If the object has already been acquired by another designer, its holder and its expected time of return are identified. "Deadlock" is rare, but nonetheless possible. Since designers can always determine who has what objects, deadlock is resolved through negotiation. Once the needed objects are at the workstations, the designer manipulates these with his design tools.

Savepoints, described in more detail below, protect transactions from loss of data due to local crashes. Since the changes are continuously being spooled to the server by a background process, most can be restored. Note that activity at the workstation can continue even though the connection to the server is broken (fortunately rare!), but this is not advisable. It exposes the designer to serious loss of data in a local crash.

When design work is complete, the transaction enters a *validation* phase. Verification programs are invoked by designers to check that the modified design data is self-consistent. If validation fails, the inconsistencies must be located and corrected. The transaction reenters the work phase, the problems are (hopefully) corrected, and validation is retried. Checking is usually distributed throughout the lifetime of the transaction, and need not only occur at the termination of design activity. The validation subsystem insures that all relevant constraints have been enforced before a transaction is permitted to enter completion.

During the *completion* phase, the mirrored copies of design objects are made available as new objects. New versions of a dependent object must be created simultaneously with a new version of the object they depend on. If a designer decides to abort his transaction, the global and local copies are destroyed, and the original objects are made available again for checkout.

To see how a designer creates a new design, consider the following example (see. Fig. 5.2). He wishes to create a new version of a microprocessor with a revised register file design. He invokes a design transaction, acquires the register file object from the Design Librarian, and revises it, creating a new version. To create the new microprocessor, he must first create a new datapath object incorporating the new register file. He acquires the datapath object, and recomposes it from its original components and the new register file. He then acquires the microprocessor object, and repeats the process. The new microprocessor design, with revised datapath and register file, is validated. The design transaction can now complete successfully with the new objects becoming a permanent part of the design database.

Design transactions make a heavy demand on disk resources. However, redundancy is unavoidable if resiliency to crashes is to be obtained. Since the server is dedicated to providing database services to the network of workstations, it should be equipped with a large number of disk devices.

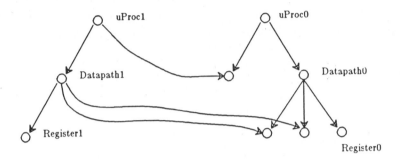

Fig. 5.2 Creating a New Design. Design objects explicitly incorporate their components. To create a new microprocessor with a new register file, you must first create a new datapath object that incorporates it. Unchanged objects can be shared across the versions of the microprocessor

5.7 Extensions to the Transaction Model

The kind of sharing supported by the design transaction model described above may be too rigid for some design interactions. The owning transaction must complete all of its design activities, including the validation of new versions of modified objects, before they can be acquired by others. *Nested subtransactions* can provide more flexible sharing, and can be easily accommodated. The transaction that originally acquires the object is the *master*. It allowed to create a *tentative version* of the object. Tentative versions need not be validated before being returned to the public database. However, they remain the responsibility of the master, and the final tentative version must pass validation before the transaction can complete. Thus, the master transaction provides a scope of consistency for the objects it has acquired. A transaction that is willing to view a partially completed object can acquire its tentative version, thus becoming a *subtransaction* of the master. Transactions that require "master" access must wait until the current master completes. When the subtransaction completes, it creates a new tentative version, which in turn can be acquired by another subtransaction or the master. When the master completes, only a tentative version that has passed validation can be added to the public database. Unvalidated tentative versions are removed.

Nested subtransactions are a mechanism for viewing a partially completed design object before its master transaction completes, and for allowing another transaction (and designer) to complete changes to the object. The first situation arises when two interacting subsystems are being designed in parallel. The two subsystems should be tested as a unit to insure that they work together properly before returning them to the Librarian. If the tests fail, the subsystems can be redesigned and the tests can be attempted again. The second situation frequently arises when a subsystem cannot be completed without the assistance of another designer, and the partial results have to be transferred to his/her workspace.

Nested subtransactions are adequate when designers can anticipate in advance what needs to be shared. They initiate an "umbrella" transaction that acquires the shareable objects and makes these available as tentative versions. It is often more natural to pass objects directly between subtransactions, rather than passing them through the public database. A centralized design server could become a bottleneck if many objects are shared. The model described above presupposes a great deal of centralization. The umbrella transaction is responsible for validating the object. However, it could be the responsibility of either subtransaction, or both, to validate a shared object. More flexible models of non-hierarchical sharing of objects that incorporate decentralized control need to be developed.

5.8 Related Work

The classic work on transactions is reported in a series of papers by Jim Gray and his colleagues at IBM Research [ESWA76, GRAY76, GRAY78, GRAY81a, GRAY81b]. Conventional transactions are simultaneously units of recovery and consistency. [GRAY78] further introduces *conversational transactions,* which remain the unit of consistency, but are not atomic for recovery purposes. Conversational transactions have nested conventional transactions as units of recovery. Committing one of these essentially defines a savepoint. A nested transaction can be undone only by a compensating transaction, which is application dependent and is not explicitly supported by the system. Design transactions, on the other hand, explicitly support savepoints, and can recover past these if desired.

Raymond Lorie's group at IBM San Jose have been extending the System-R relational database system for engineering applications [HASK82, LORI83]. Their transaction model is based on Gray's conversational transactions. Once data is extracted from the public database into a workspace, it no longer receives support from the database system. Savepoints copy workspace data back into the public database. Thus, only recovery to savepoint consistent states are supported. Version control and consistency maintenance are not considered to be functions of the design database system, and are not supported. [KIM83] extends the transaction model along the lines of the preceding section.

Some researchers have advocated a more radical view of design transactions. Eastman [EAST81, KUTA83] views a design transaction as a flow of information through a collection of design applications, some adding data to the design description, and others which check its correctness. While appropriate for well-understood design domains, such as building construction, the approach is too restrictive for the still developing VLSI design domain. The design process cannot be modelled a priori as a flow through design tools. While some dependencies do exist, e. g., a PLA layout can be machine generated from the Boolean expressions that describe it, most of the design is only loosely related. For example, a system architect defines a functional subsystem, a logic designer maps this into a schematic, and a layout artist creates the layout.

While not immediately applicable to VLSI design, Eastman's transactions do capture some of the complicated consistency constraints of design data, i. e., dependencies among different representations of the design. Similarly, Neumann [NEUM82, NEUM83] defines design database consistency in terms of independent and dependent representations of design objects. A design object is consistent if all dependent representations have been derived from the most up-to-date versions of independent representations. Unfortuately, this assumes that a strict dependence hierarchy can be defined among design representations. This is not the case for VLSI circuit representations. For example, layouts are not always created from circuit descriptions. In some cases, the circuit description is extracted from the layout.

The transaction model presented in this chapter resembles that of [HASK82, KIM83, LORI83], but with several subtle differences. When a design transaction completes, new versions of modified objects are added to the library rather than overwriting its current contents. Version creation is intimately related to transaction management. Recovering design objects to a savepoint consistent state is inadequate. Mechanisms are needed to support continuous recovery. If transactions are to remain the unit of consistency, mechanisms to support design validation should be incorporated within design transactions.

6 Design Management System Architecture

6.1 Introduction

A central theme of this book is that an integrated design database is needed before a collection of design tools can be forged into a truly integrated design system. In this chapter, we present an architecture for a prototype design management system. Recall that a design management system is much more than a database system: it is responsible for choosing the data structure for representing the design and for providing an appropriate interface to this structure for design tools. The structures of the system's storage, recovery, concurrency control, and design validation subsystems are described. Special applications programs that manipulate the structure of the design, a *design browser* and a *system assembler,* are also discussed.

6.2 System Architecture

A designer naturally views the heart of a design system to be the design tools with which he interacts directly. From the perspective of a design tool builder, the job of implementing a new design tool could be simplified if some of their services were available as stand-alone subsystems. The primary candidates are subsystems for screen (menus and windows) and data management. Database systems arose in response to a proliferation of different file management code within applications. To promote sharing among these, their hand-crafted file management routines were replaced with a single standardized interface to shared data. Stand-alone menu/window packages and report writers/forms applications systems are now available. A *design management system* can provide a similar service for the data management aspects of design tools service.

The overall system architecture appears in Fig. 6.1. We discuss each of its components in turn. The *Storage Component* stores design data on disk and guarantees that updates are atomic. A conventional database system can be (and has been!) used as a storage component, although a suitably extended file system could be used as well.

The *Object System* maps the design data, viewed as a collection of interrelated objects, into the files and structures supported by the storage component. A reliable object-oriented file system can provide the facilities of both the Storage Component and the Object System. However, if the system is to continue to support existing design tools, the object system should be able to appear as a conventional hierarchical file system.

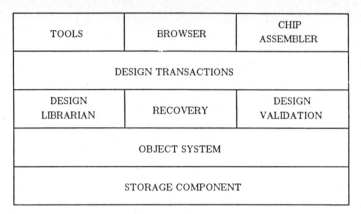

Fig. 6.1 System Architecture. The foundation of the system is the STORAGE COMPONENT. The OBJECT SYSTEM implements an object-oriented design description based on the primitives supported by the STORAGE COMPONENT. The DESIGN TRANSACTION SYSTEM is founded on (1) the DESIGN LIBRARIAN, which controls sharing by restricting access to objects; (2) the RECOVERY COMPONENT, which spools changes from workstations to database servers; and (3) the DESIGN VALIDATION COMPONENT, which assists in determining the correctness of an object before it is made public. Design tools are constructed on top of this

The *recovery subsystem* collects incremental changes made to objects, ensuring that they can be reconstructed after a crash. To protect workstations from data loss, changes are continuously spooled to the server. *Savepoints* invoked by the designer force in-progress changes to be saved, thus guaranteeing that the object can be reconstructed to that point.

The *Design Librarian* controls shared access to design objects. Although many designers can browse an object, only one at a time is allowed to create a new version of it. The Design Librarian maintains information about who currently holds objects and when these are expected to be returned to the design repository. This enables the recovery subsystem to reconstruct the workstation's disk after a hard crash.

The *design validation subsystem* interprets dependencies among design data to identify those portions that may be affected by a change. Some simple consistency checking can be handled automatically, particularly for representations understood by the subsystem. More complicated validation, e. g., verifying equivalence across representations, requires designer intervention. The component maintains an audit trail of what validation activities have been performed by the designers, and provides analysis tools to assist the design team in determining what remains to be validated as the design reaches its release state.

Encompassing the design librarian, recovery, and design validation subsystems, the *design transaction component* ensures that designers create new consistent versions of design data with their tools. *Design transactions* (introduced in the previous chapter) bring data to the workstation or the designer's workspace, returning it as new versions when done. Designers acquire access rights to the appropriate portions of the design from the Design Librarian. Successfully aquired objects are then transferred to the workstation. The recovery subsystem guarantees that all but perhaps the most recent changes can survive system crashes. The validation subsystem determines what is affected by the changes, identifying what must be reverified. A design transaction cannot complete successfully until the design is once again consistent.

Design tools manipulate the data through operations supported by the object system at workstations. The familiar design tools create or verify the representational portions of the design description. Other tools are needed to manipulate the structural aspects of the design. The *Browser/Chip Assembler* is the interactive interface to the design data management system. Designers use the Browser to view the complex data structure describing the design. They use the Chip Assembler to manually construct the data structure from the pieces created by individual design tools.

Each of these subsystems is described in greater detail in the following subsections.

6.2.1 Storage Component

The Storage Component manages design data on secondary storage. Although a conventional database system can serve as a storage component (see, for example, [HASK82a]), a transaction-based file system can also provide the required facilities. For example, the popular UNIX file system must be extended with atomic update capabilities before it can be used as a design storage component.

The Storage Component is a reliable archive for *design files* residing on the remote database server. The notion of design objects is implemented at a higher level of the system. When a designer creates a new version of an object, he first acquires exclusive access to it from the Design Librarian, a client of the storage component. Unlike a conventional database system, the Storage Component does not need special concurrency control mechanisms. The Design Librarian processes requests one at a time, thus serializing access to the Storage Component. A copy of the original file, created at the server, is simultaneously transmitted to the designer's workstation. Once again, unlike a conventional database system, complex storage structures are unnecessary, since the Storage Component reads and writes whole files. Once at the workstation, the files are processed by design tools that typically read the files into virtual memory, manipulate the data in virtual memory data structures, and write them back when done. The function of the Storage Component naturally extends (although non-trivially!) to the management of this in-memory data.

The Storage Component and the recovery system must interact closely. *Save actions* are supported by the recovery system, which must in turn be supported by the Storage Component. Changes made to the files at the workstations are kept in special logs. At a savepoint, the logs are transmitted back to the storage component to be merged into the server's copy of the file. The Storage Component must guarantee that the merge operation is atomic, ensuring that files are never left with partially merged changes. The merge can be implemented as a conventional database transaction, if such transactions are supported by the Storage Component. Otherwise, the atomic merge can be implemented by creating a temporary copy, merging the changes, renaming the temporary file, and removing the original. The merge can be restarted if a crash occurs. While the merge may appear to be time consuming, it can be overlapped with continued activity at the workstation. Individual design files are relatively small and have a simple structure, so the time required to do a merge need not be excessive.

6.2.2 Object System

Design objects are implemented as files containing representational primitives and design management information, including the object's type, the names of other objects that contain this object as a component, the names of the component objects contained within this object as a component with a description of how they are to be composed, and the object's interface specification. The data model supported by the Object System was presented in Chap. 4.

The Object System maps the abstract notion of an "object" into the data structures describing it. It maintains directories that map object names and version numbers into files. Such information can be maintained within conventional database index structures if a general purpose database management system is used for the Storage Component.

If a file system is used for the Storage Component, then design objects can be stored either as a single monolithic file, with combined representation and structural information, or as separate files. For representation types not known to the system, data are stored in a separate file referenced from within the design object file. The object system can pass this "raw" representation to existing design tools, thus appearing to them as a conventional file system. The system uses the structural information (e. g., the design hierarchy, the equivalence links across representations, etc.) for validation and browsing while traversing the complex structure of the design.

6.2.3 Design Librarian

The Librarian coordinates all access to shared design data, making design objects available to workstations. A designer acquires only one object at a time. The browser is used to navigate the hierarchies to find the object of interest. Checking out an object for update gives the designer the exclusive right to create its new version. The mirrored copies of design files, supported by the recovery subsystem on top of the Storage Component, become the new object versions when the changes are committed (see Fig. 6.2).

Several designers could simultaneously create new versions, but such a proliferation of versions is undesirable for effective project management. By employing check-out locks, the Design Librarian guarantees that only one in-progress version of an object can exist at a time. Some designers feel that this limitation is too restrictive. They would prefer that the Design Librarian simply warn them when they check-out a version of an object when there is already one in-progress. The second designer could then choose to abort his check-out, or to continue as normal. This complicates the Librarian's bookkeeping (e. g., what is the "latest created version" in this case), but is not difficult to support.

Suppose we continue with the "one in-progress version" restriction. In the rare instance of one designer holding an object that another designer needs, the designer who has it is permitted to complete its revision, creating a new consistent version of the object. Afterwards, the requesting designer can acquire it, creating his new version based on any one of its previous versions. That is, he is not restricted to creating a new version of the most up-to-date version, although this situation will be the most com-

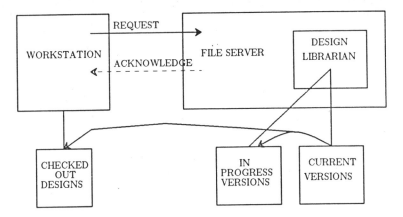

Fig. 6.2 Design Check-Out. Objects are checked out by directing a request to the Design Librarian running on the file server. The request is acknowledged by transferring the desired object to the workstation. A back-up "in-progress version" of the object is simultaneously created at the server

mon. It is still possible to create an arbitrary branching tree structure of versions of the same object. The history of which versions have been derived from which is maintained by the object that aggregates all versions of the same object.

Note how the design environment's requirement to support versions can simplify the task of the Design Librarian. They provide a flexible vehicle for managing concurrent access to design objects. Previous versions can be browsed without regard to in-progress update activity. Simple locking protocols are sufficient to protect against the proliferation of versions.

6.2.4 Recovery Subsystem

The Storage Component guarantees that the underlying structures on disk can be updated atomically. The Recovery Manager provides recovery support between the workstations and database servers. Its goal is to insure that as much data as possible survives a system crash. It accomplishes this by maintaining multiple copies with different failure modes: (1) an up-to-date copy at the workstation, called the *local work file;* (2) a file of accumulated changes since the last savepoint, kept at the workstation, and called the *local change file;* (3) a file that is a snapshot of the local work file as of the last savepoint, maintained at the file server, and called the *mirror file;* (4) a file that contains spooled changes since the last savepoint, kept at the file server, and called the *global change file;* and (5) the *redo log,* holding all the changes since the last archive, and residing at the file server (see Fig. 6.3). The local and global change files, containing the differences between the local work file and its mirrored copy, protect data from being lost during workstation crashes. The redo log, containing the difference between the mirrored file and the original file, protects against losses during file server failure.

The recovery manager supports *savepoints.* At a savepoint, data and change file buffers are forced to disk by the workstation's buffer manager. The local change file is copied to the database server and atomically appended to the global change file. A background process copies the local change file entries to the global change file, providing a *continuous recovery capability* that guards against local crashes and reduces the latency of a save. To commit the changes, the storage component atomically merges the global change file into the mirrored file and appends the global change file to the redo log. Space on the local disk is reclaimed as local change entries are copied back.

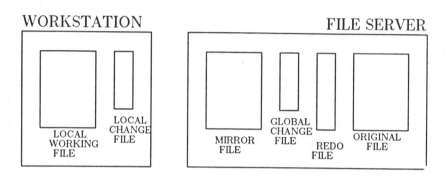

Fig. 6.3   Files Associated with a Checked Out Object. An object is implemented by a file. To provide continuous recovery of changes made to this file, several copies of it are maintained throughout the system. The local files are at the workstation, while the global files are at the server

A *soft workstation crash* leaves its disk intact. The local change file is undisturbed, and only workstation buffers are lost. In this case, recovery proceeds as follows. Recovery to the last savepoint is achieved by simply copying the mirror file from the server back to the workstation. An ever more up-to-date version can be recreated by merging the global and local change files into the mirror file before it is copied back. Some updates may be lost, because they were in local buffers, but updates written to disk since the last save can be restored.

A *hard crash* loses data on the workstation's disk. Both the in-progress version on the workstation's disk as well as the local change file are assumed to be lost. As above, copying the mirrored file back to the workstation restores the file to its last saved state. Alternatively, copying back the merged mirrored and global change files also restores the file to the last state known to the file server. The recovery subsystem determines what objects should be restored by asking the Design Librarian.

The database server employs more conventional techniques to ensure that its files are durable. The system is made resilient to soft crashes because of the atomic actions supported by the storage component. Frequent archival dumps of the original files provide resiliency to hard crashes. The mirrored files can be restored to the last savepoint state by merging the redo log into the archival copy of the file. However, the recovery subsystem guarantees only that objects can be restored to their last savepoint, although every effort is made to restore beyond that. This is achieved by duplexing the redo log but not the change files. Spooled changes not yet saved are lost, since they are not normally logged. If the server loses the accumulated changes, it can always acquire the most up-to-date version of the file from the workstation.

6.2.5 Validation Subsystem

The validation subsystem helps designers track what must be revalidated after a design change. Since validating the fully instantiated design is prohibitively expensive, the design's structure must be exploited to keep the validation effort reasonable.

A design is self-consistent if *conformance, composition,* and *equivalence constraints* are in force: (1) an object's implementation satisfies its interface (*conformance*); (2) the composition of component objects is well-formed (*composition*); and (3) objects specified as equivalent across representations are shown to be equivalent (*equivalence*). Design transactions cannot complete until all constraints are satisfied.

A completely automatic approach for design validation is not yet obtainable. Designers will continue to need to step in to execute the appropriate validation tools and to interpret their results. However, the Validation Subsystem can provide mechanisms to assist in the validation effort. For example, it can provide an audit capability. Designers report who they are, what constraints they have validated, and what tools have been used for the validation. Once in machine readable form, the audit trail can be analyzed to provide accountability and to indicate which portions of the design still have to be validated. If a design failure is traced to a particular design object, the designer responsible for its "validation" can at least be identified.

By making structural information explicit, the object model greatly assists in the enforcement of these constraints. Checking conformance constraints requires the development of new tools for interface extraction. Some parts of the interface – such as the location, type, and drive capability of ports – are easy to extract. However, performance aspects such as delay information can be determined only through simulations performed by the design team, who then reports the results to the subsystem. Note that these aspects of the interface are explicitly recorded in the object's interface specification.

The checking of composition constraints is well supported by the Object Model's notions of ports and interconnections. A composite object is well-formed if the interconnected ports of its components are type- and direction-compatible. That is, component ports connected to the composite object's ports must be type-compatible and have the same direction. For many representation types, such as layout descriptions, this kind of information is implicit. An interface description makes these explicit.

The Object Model explicitly incorporates structures for identifying equivalent objects across representations. These have been included to reduce the effort needed to validate equivalence across representations. Suppose we must show equivalence between the layout and transistor representation of a design. Normally, equivalence is checked by first extracting the layout into a transistor description, and then providing the same stimulus to simulations of both the original transistor description and the extracted description. Ideally, only the changed portion of the layout description and parts of the design that depend on them should be extracted and simmlated. If two objects in different representations are already equivalent because they are connected by an equivalency object, the transistor object can be substituted for its counterpart in the extracted tranistor description without further analysis. A multilevel simulator can simulate the object at this level without fully instantiating it.

A general mechanism to assist in keeping track of the validation status of the design uses the audit trail and is based on a set of parameterizable programs for its analysis.

The parameterization includes: (1) dependency information; (2) equivalence constraints across representations; and (3) check-in scripts. For example, the analysis program can be informed that PLA layout objects are dependent on PLA Boolean equation objects, and are derived through a certain sequence of execution of design tools, such as a Boolean equation minimizer and a PLA generator. The identification of which objects are PLA layouts is part of the information stored in the design object descriptions. Entries for the program executions and the creation times of the objects must be consistent with the dependency specification. For example, if (1) the last update time of the PLA layout object is the same as when the PLA generator was run, and not later, and (2) the last update time of the PLA Boolean expression object precedes this time, then the objects are known to be equivalent. Such equivalency constraints can be actively enforced by regenerating dependent objects on request.

Again, the Object Model assists in the validation effort by making explicit the equivalences across representations. Validation sequences for checking that two objects in different representations are equivalent is specified to the analysis program. It checks that the appropriate sequence exists in the audit trail.

The analysis tool also supports check-in scripts. Validation sequences for objects of a particular representation (e. g., "every layout object must be checked by the design rule checker"), and for equivalent objects across representations (e. g., "every layout object must be extracted into transistors and compared with its equivalent transistor object") can also be defined, and checked by the tool.

6.2.6 In-Memory Databases

6.2.6.1 Introduction

It is natural to think about extending the design management system into the management of in-memory data structures. However, there are some difficulties. The internal and external representations of design data are often quite different. When a tool like a circuit layout editor begins execution, it first reads in a character file describing the geometries of the design. It then construct an in-memory data structure that contains this information in a form more suitable for rapid display on a graphics screen. Over time, the tool manipulates the internal form in response to the designer's commands. When a session is complete, the tool reformats the internal form into the character oriented external form, and writes this back to disk.

The problem with the preceeding scenario is that once the design description leaves the file system, it can no longer enjoy the nice recovery features that we have come to expect. A system crash will lose the contents of memory, and all changes since the last "save" will be lost. Even if the file system is replaced by a database system, the latter's recovery mechanisms do not extend to the data structures formed by the design tool. The recovery mechanism described in Sect. 6.2.4 only applies to changes made at the workstation to design files, not to the internal structures created and manipulated by the design tool. The question becomes how to extend database facilities to the management of in-memory data.

Throughout this subsection, we will use a very simple model of the mapping between internal and external formats of data. We shall assume that design data is represented on-disk as character strings of nested S-expressions. This is precisely the external format we have chosen for the Object Model specification in Chap. 3. The internal representation is linked lists of CONS cells in memory. *Atoms* are atomic objects, usually character strings, which have no pointers, while *CONS cells* contain two pointers to atoms or other CONS cells. A list is an interconnection of CONS cells such that each lefthand pointer points to the list element while the righthand pointer points to the CONS cell for the next list element. The CONS cell for the last element of the list is denoted by a NIL righthand pointer. When an S-expression file is read into memory, a complex web is formed of atoms and lists pointing to atoms and other lists (see Fig. 6.4). The mapping is straightforward: individual words are mapped into atoms while each pair of parentheses in mapped into a list.

6.2.6.2 Building In-Memory Structures: Complex Object Mapping

In Chap. 3, we described how complex objects are structured in extended System-R. The user interface to the system has been extended to allow the application programmer to acquire an entire complex object at a time. First, the root tuple of the complex object is identified and read into in-memory buffers. Then its immediate descendents in the COMP-OF hierarchy are accessed and are linked to the root tuple. The process continues with descendants of these tuples, until all tuples connected to the root through a transtive closure of COMP-OF linkages are exhausted.

Given that the COMP-OF relationships among tuples are repesented by placing the identifier of the parent tuple in its related child tuples, it would appear that building such a structure is potentially very expensive because of the repetitive join operations involved. However, this need not be the case, and a different approach has been taken by the designers of the the System-R extensions. While the user thinks of a COMP-OF field as containing the logical identifier of the parent, this information is encoded in a more complicated way to achieve better access efficiency. A special MAP table is associated with each complex object instance, with a row for each constituent tuple of the complex object. A row tabulates a tuple's logical identifier (a character string that is unique across the entire database) with its physical tuple or record identifier (its physical address within the database). The latter can be used to directly access the tuple on disk. A COMP-OF field is actually implemented as two columns: one to identify the root tuple and, thus, the complex object instance and its associated MAP table, and the other to specify an integer index into the MAP table to find the tuple id of the referenced tuple.

The relationship among a parent tuple and its children are implemented by threading these references through the tuples in the following way. The parent tuple contains a reference to the first child tuple of a given tuple type. A thread links together all the children tuples of the same parent. If these children are in turn the parents of other tuples, each will contain a new link to its first child tuple. A reference can be followed by at worst two disk operations: one to index into the MAP table, and a second to actually fetch the referenced tuple.

It is easy to see how an efficient in-memory data structure can be constructed given this on-disk representation. When an object is accessed, the MAP table is read into memory, and extended with a third column for in-memory addresses. As the tuples are brought into memory, the table entries are updated with their in-memory locations. The in-memory tuples continue to contain logical references, but these can be dereferenced to physical addresses very quickly through the MAP table. As tuples are added

```
(INTERFACE
    (NAME nand)
    (TYPE primitive)
    (PORTS
        (in1 INPUT)
        (in2 INPUT)
        (out OUTPUT)
    )
)
```

(a) External Format

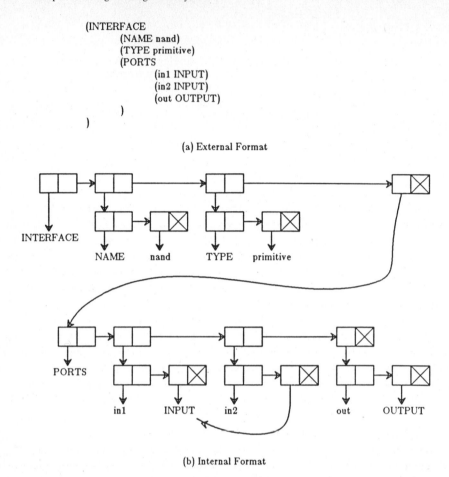

(b) Internal Format

Fig. 6.4 Internal and External Representations of a Design. Specification for a NAND Gate. (A)shows a LISP-like S-expression description of a NAND gate, while (b)shows how it could be implemented by interconnected "cons" cells

to the object in memory, new entries are made in the MAP table. As tuples are added to the object in memory, new entries are made in the MAP table in a straightforward way. When the object is written back to disk, the MAP table is used to determine where to find the physical records that should be written to disk.

The advantage of the approach is that physical memory addresses are never stored in the tuples, preserving their location independence. New tuples can be added to the object in a location transparent way. The disadvantages are related to the complexity of managing the MAP tables in memory. There is one such table for each active object, and one entry in the table for each tuple in the object. This could take up a very large amount of the available buffer memory, unless a limit is placed on the number and size of active objects. Further, we have not described how to map between the object identifier and its associated MAP table. Another in-memory table must be maintained for this purpose.

6.2.6.3 In-Memory Recovery

Recovery mechanisms are needed to support the manipulation of design data stored in main memory data structures. Such structures are normally lost in a system crash. A recovery technique based on *operational logging* can be used.

A well known recovery mechanism is the *write ahead log protocol* [GRAY78]. In addition to the collection of records stored on non-volatile media, which represent the current state of the database, a database system also maintains an incremental append-only log. The log is a long sequence of records that document the changes made to the database since the last archive. Only the most recent log entries are maintained on-line; the rest are written to tape and eventually discarded.

Changes to the database are normally made in-place. A transaction either completes successfully, and is committed, or fails to complete, and is aborted. If the transaction commits, the recovery manager insures that the effects of the transaction are durable. If it aborts, the recovery manager removes any changes that have already been written to disk. Log records should be written to disk *ahead* of pages containing their associated update record, to make it possible to redo committed updates and undo uncommitted updates by having the log safely on disk.

The recovery mechanism is complicated by the buffer pool between the database system and the disk. Writes (DO operations) issued by transactions change records in the buffer pool, and are not written immediately to disk, although old pages are periodically written out to disk because of buffer replacement. If a transaction has committed before the crash, the recovery manager must REDO its actions that were reflected in the buffer pool but which had not been written to disk. If a transaction is in progress at the time of the crash, or it is aborted by the user (or by the system in deadlock situations), the recovery manager should UNDO its effects from the database.

Any write to the database creates a new entry in the log. The most essential part of the entry are the old and new values of the changed record. The recovery manager UNDOes the effects of an update by replacing an object's current value with the old value in the log. It REDOes the update by replacing the current value with the new value stored in the log. Additional log data identifies the page that was changed, and chains together entries made by writes of the same transaction.

If the database does not have a uniform record structure, several difficulties arise in applying the traditional "Old Value/New Value" log approach for recovery. Relationships among database records are usually represented by symbolic values, data structures built in-memory represent relationships by physical pointers. For example, a CONS cell of an S-expression consists of two pointers to either atoms or other CONS cells. Logging old values and new values of these pointers is not sufficient. Every time the data structure is reconstructed in memory, the pointer values could change because the data structure is constructed in a different region of memory.

In conventional database systems, each record has an associated record identifier, which locates it within a specific page of the database. Changes to the record in the buffer pool are correlated with its on-disk version by placing this identifier in the log entries. There does not appear to be a similar way to correlate changes to in-memory S-expression pointers with their ondisk representation.

Rather than viewing the log as a sequence of old/new values of records, we view it as a sequence of operations done on the data. In the event of a crash, we read in an old state of the database and "replay" all the operations recorded in the log since this state of the database was created. This is easy to do in the design environment, because the original design file version is maintained while the next version is being created. Conventional systems overwrite the database's contents, and then require special mechanisms to undo those writes.

Figure 6.5 illustrates before and after images of the in-memory list structure that converts the 2-input NAND gate into a 3-input NAND gate: "add (in3 INPUT) to PORTS". This operation is broken down into the following individual update operations.

(1) Create CONS cell C1.
(2) Create CONS cell C2.
(3) Create CONS cell C3.
(4) Create ATOM "in3".
(5) Set C1's right pointer equal to the right pointer of CONS cell (Cp) whose left pointer points to PORTS.
(6) Make Cp's right pointer point to C1.
(7) Make C1's right pointer point to C2.
(8) Make C2's left pointer point to in3.
(9) Make C2's right pointer point to C3.
(10) Make C3's left pointer point to INPUT.
(11) Make C3's right pointer point to NIL.

Rather than log one change record for each of the above writes (as already pointed out, there are difficulties in recording old/new values of pointers), we record the operation "add (in3 INPUT) to PORTS". This saves a significant amount of space in the log.

The above sequence of writes has to be processed as an atomic operation. Since we do not log each of the individual writes, there is insufficient information to undo the writes of an incomplete operation. Therefore, our log is REDO only, rather than the conventional UNDO/REDO. At restart, we keep around the previous state of the object until the operation is completely processed.

In a typical transaction processing environment, a sequence of writes, such as above, is made atomic by bundling it into a transaction. With an operational approach, we avoid much of the transaction management overhead by logging high level operations rather than the individual writes that implement them. Of course, it takes longer to actually recover the database, but this is hopefully the exception event! Instead of calling the transaction manager, the application program is responsible for recording its update operations in the log. Updates to objects are recoverable as soon as the log buffer is written out to stable media.

If individual operations (such as "add x to y") are like transactions, then successful completion of an operation is similar to end-transaction. The log should be forced out to disk to commit the operation. To save I/O traffic, we relax this requirement, and we only force the log buffer to disk when it becomes full. As a result, some successful operations might not be recoverable, but this is acceptable in an interactive update environment. Of course, at a savepoint, the entire log buffer is forced to disk.

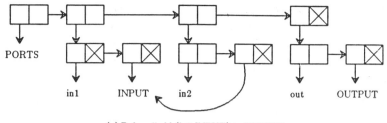

(a) Before "add (in3 INPUT) to PORTS"

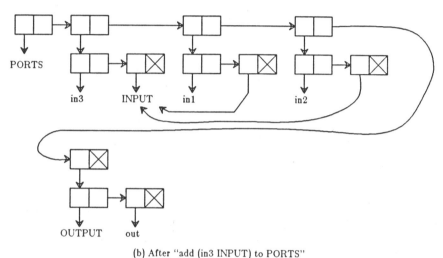

(b) After "add (in3 INPUT) to PORTS"

Fig. 6.5 Adding an Input to a NAND Gate. The figure shows the changes made to the internal data structure to expand the description from a two-input NAND gate to a three-input NAND gate

Although the operational logging idea is simple, its efficient implementation is not straightforward. For example, in the sample structure given above, we do not know where the character string "PORTS" resides on disk. We have two choices when restarting from a crash. One is to search the file that stores the above S-expression for a match with "PORTS" and then append the list ("in3 INPUT)". Although efficient text searching algorithms do exist, in the worst case we must process the entire file. The second choice is to read the entire S-expression into memory, construct its in-memory data structure, reapply the operation, and write the S-expression back to disk. Either approach is expensive when compared to the more conventional method available for record-based databases. The record-id can be used to identify the precise location of the record to be restored. The page on which it resides is reread, the new value of the record replaces the current value on the page, and the page is written back to disk.

Similarly, checkpointing is more expensive because it requires internal data formats to be translated into external formats, and an object's pages must be completely rewritten rather than just the changed pages.

6.2.7 Version and Configuration Management

6.2.7.1 Introduction

It is not surprising that designs evolve over time. New versions of system components are created and are incorporated in new configurations of the design. In this subsection, we will discuss methods for organizing the versions and configurations of a design. We have already discussed something about version control in the presentation of the Object Model in Chap. 4. Special *version* and *alternative* objects were introduced for organizing the collection of objects that describe a design across their versions. Configuraton information was embedded in the composition structure of the objects: if a new version of the ALU is created, it must be explicitly "configured" into a corresponding new version of the datapath object that contains it.

6.2.7.2 Design Administration

For the purposes of version control, a *design* is represented as a synchronized collection of versions of related design files (think of a design file as containing the description of an individual design object). *Design files* are homogeneous collections of record instances storing information about the design. A *design version* specifies which versions of individual design files constitute a particular version of the object being design. In the Object Model, there really is no difference between a design version and a version of a design file, since composition information is embedded within the file. However, it is possible to factor out such information into a separate *configuration file,* which describes how the design is configured from collection of design object version (see Fig. 6.6). Versions of the configuration file represent different versions of the design.

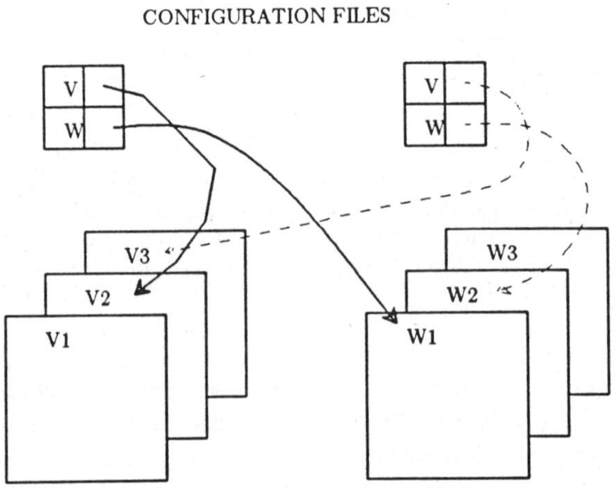

Fig. 6.6 Configuration Files and Design Versions. Configuration files tie together versions of individual design files to form versions of a design

To illustrate how versions and alternatives can arise in the design environment, consider the following system life cycle scenario (see Fig. 6.7).

Each design file has its own *administrator,* who is responsible for its contents. Other designers can make their own private copies which may be updated. We call these *alternatives.* Only the file's administrator can incorporate the changes into the file.

A new *in-progress* file version is created from its latest released version. Only its administrator can update the in-progress version. Other designers can create alternative versions of the file. Changes to alternative designs are not reflected in the in-progress version until explicitly merged into the latter by the administrator.

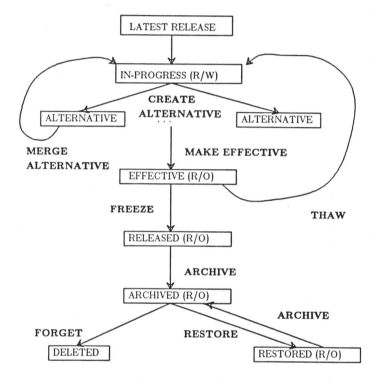

Fig. 6.7 Design Life Cycle Scenario. A possible model for a design life cycle is shown. A previous release is made in-progress and can receive new updates. Various alternatives can be created in parallel, and one is selected to become effective. If the effective version is valid, it can move on to become a new release. Old releases are eventually moved to archive, and eventually discarded

Effective versions are read-only and represent a pre-release version of a design file, suitable for testing before actual release. The administrator merges alternatives into the in-progress version before making it effective. Once effective, a file can be updated only if it is first thawed back into an in-progress version. Otherwise, it can be frozen into a *released* version.

Releases can be archived and later restored. Ancient versions are removed from the archive by being forgotten. Once forgotten, a version can never be reaccessed. Usually direct access is required for the latest release, all effective versions created since the last release, and the current in-progress version. Thus, it is convenient to have several versions of each design file available on-line.

Effective versions are *design checkpoints*. Design data must pass a battery of verification checks for self-conistency before they can be released. Effective versions are firewalls, providing a capability for backing up a file to a previous consistent version known to have passed these tests.

Alternative versions, in the sense we have used them here, are different from other versions. A designer often attempts several solutions to a design problem. For example, he may design two circuits that perform the same function, one consuming less area and another less power. The design system should aid him in carrying forward multiple alternative designs in parallel. An alternative can be combined with the current in-progress version to form a new in-progress version. Because alternative designs are tentative, changes do not affect the current working copy of a design file until it is explicitly incorporated into the in-progress version.

[BOBR82] contains a description of how to implement a version and alternative mechanism that could support the design life cycle described above – although their concern is primarily how to manage versions of artificial intelligence knowledge bases. A design file contains a heterogeneous collections of objects, although we can think of these as a collection of S-expressions if we wish. Versions and alternatives are implemented by dividing the file into *layers*. The initial layer contains the original version of the file, the second layer contains new objects added to the file and new copies of old objects that have changed in creating the next version, etc. Each object is uniquely identified, making it possible to reference different copies of the same object across layers.

Layers provide a very flexible mechanism for organizing design versions. They need not be searched from newest to oldest when looking for an object. Layers can be reordered or omitted from the search order. The first definition of an object found in the specified order is its version within that *environment*. An environment, existing for each designer on each design file of interest, is an index structure that maps unique identifiers into objects, taking account of a designer's desired order of the file's layers. An environment can also contain layers from the designer's private files, thus making it easy to form multiple alternatives of the design file.

For more efficient usage of storage, a design file's administrator can summarize a design file, replacing the file with the most recent versions of all objects in the file. Since this could disrupt the view of dsigners who have created alternatives of the file, they have the option of freezing the file, which amounts to making copies of all the layers they refer to in their own environments. Old copies can be discarded by thawing the file. Before summarization takes place, presumably at a consistent point in a design file's evolution, its administrator snapshots the file into an archive.

We have described the implementation of versions of a design file by using layers, but a design version is actually a collection of design file versions. It is specified by a set of environments over the constituent design files, including the design hierarchy file, composition files, representation files, and interface files if these are stored separately. Thus it is possible to have a version of a microprocessor with a new ALU and an old barrel shifter.

6.2.8 Design Applications

6.2.8.1 Design Browser/Chip Assembler

The Browser and Chip Assembler are the interactive interfaces to the design management system (see [KATZ83b, MUDG80, MUDG81]). The Browser allows a designer to navigate through the complex structures describing a design, making extensive use of graphics to present the data structure, and menus to direct the navigation.

Part of the motivation for interface descriptions is to provide on-line design documentation. Browsing through object versions provides insight into how it has evolved over time. Alternatives within design libraries can be scanned to find a particular implementation of a function suitable for incorporation into a new design. A design can be browsed within a representation, or across representations, at any level of composite object or at the fully instantiated level of primitive objects. Several ways of browsing the design can be underway at the same time. Each browse takes place within its own window.

The Browser assists the designer in finding objects with particular attributes. Index objects group together objects with similar attributes. For example, the ALU index object is composed of all generic ALU objects stored in the design database. The ALU index object can be further included in a datapath index object, etc., creating a hierarchy of indices. Index objects provide entry points into the design database.

The design database is likely to contain more objects than can fit comfortably on the screen, even if we represent these as simple nodes with edges to denote the composed-of/is-a-component-of relationships among objects. Therefore only design objects within a particular context are represented on the screen, e. g., the layout "plane" of the design or all objects aggregated together by the same version object.

Another possible viewpoint into the design is to present the representation hierarchy looking down upon it rather than sideways. The composition relationships are denoted not by directed edges, but by containing design polygons within other design polygons. This is commonly used by geometry editors to represent subcells within cells, but the approach can be used for any representation type. It is just another way to view the design hierachy.

The Chip Assembler takes data created by synthesis tools, such as graphic editors and datapath generators, and packages them as objects for inclusion within the design database. In conjunction with the Browser, it provides an interactive front-end for constructing the design data structure. It supports the interactive (1) composition of objects from more primitive objects; (2) specification of object interfaces; (3) identification of equivalences across representations; (4) construction of design configurations through versions and alternatives objects; and (5) creation and manipulation of index objects.

7 Conclusions

7.1 Research Directions

Design databases, as viewed from a database systems perspective, are still a relatively unexplored territory. In this subsection, we discuss a number of database techniques that could apply to data management problems in the design environment if they are suitably generalized.

Specifying design constraints in terms of database consistency constraints and triggers:

While we have argued that database consistency mechanisms are currently inadequate for handling design constraints, they conceivably could be generalized to do so. A database integrity constraint specifies a static condition that must be true before a transaction is allowed to commit. They are not normally concerned with constraining the sequence of operations through which data is changed. These transitional constraints, or "scripts", appear to be very important for design data conistency. Extensions to conventional database consistency mechanisms for implementing transitional constraints may also prove useful in the conventional database environment. While we have described an off-line method for checking scripts against an audit trail of events in Chap. 6, an incremental and on-line approach would be more desirable. How to incrementally match a set of scripts against the history of design events remains to be developed.

Database triggers are user defined procedural attachments to the database system that are invoked when specified events are detected. By extending the database system to in-memory data, database triggers could be made to provide better consistency support for design applications. For example, design rule checking can be made more efficient when data structures are employed to capture the adjacency of layout objects. A trigger mechanism could be used to implement continuous design rule checking if it is invoked for every update operation, and if it were parameterized with the small number of affected geometries that need to be checked. A mechanism needs to be developed in which procedures, such as the DRC, can be embedded within the memory manipulation interface of the system.

The design database as a distributed as well as dispersed database:

We have proposed a model of a design database in which data is stored in a central server and dispersed to workstations on demand. Once high performance disks are available on the workstations, a decentralized, distributed database organization may be more appropriate. The Librarian service provided by a single machine may prove to be a performance bottleneck. A distributed database approach provides a more uniform interface, since it can support multiple servers and data distributed on workstation disks without the designer needing to know the precise location of data. However, for adequate performance, the data will have to be staged close to where it is being used. The nested transaction model of Chap. 5 could be implemented more easily if data an workstation disks could be accessed directly by other workstations. In the described scheme, tentative versions must be shipped to the Design Librarian before being forwarded on the subtransaction.

Decentralized algorithms for acquiring objects and for providing recovery to workstations need to be developed:

Workstation data still needs to be backed up on file server machines to provide continuous recovery. Rather than providing a general file service, these machines serve as stable storage. Since general query processing need not be supported, many aspects of the implementation can be simplified. However, browsing capabilities need to be supported, and this must be accomplished with high performance. Efficient ways of keeping workstations informed about the structure of the design as it changes need to be developed.

Optimistic methods can be used to support to design checkout and return:

Optimistic concurrency control mechanisms are based on the assumption that interference among transactions are rare. Transactions acquire copies of the data they read or write without setting locks. Concurrent updates are directed to these copies. At transaction end, a validation phase determines whether a transaction's updates have conflicted with another completed transaction. If not, its changes are merged into the public copy of the database. Otherwise, the transaction is aborted.

The idea may also have some application in the design envrionment. Since actual designer interference is rare, most transactions should complete successfully. One difference is that conflicts should never lead to a transaction abort. Alternatively, mechanisms could be developed for merging the changes with the help of the designers. Some techniques have already been developed for distributed multi-copy databases where the copies have evolved independently because of a network partition [PARK81, PARK82].

View mechanisms can be generalized for maintaining consistent alternative representations of the design database:

Requirements for design data often conflict. While additional structure for enhanced consistency maintenance is desirable, certain high performance applications need specially tuned representations, with a minimum of extraneous data. Graphical database are an example of the latter.

Two approaches appear feasible. Either the relevant portion of the database is extracted and reformatted according to the needs of the application, or two representations of the same object are maintained in parallel. The latter approach is preferred when the extraction time is a significant fraction of the time the designer interacts with the special representation. The solution strongly resembles an implementation of concrete database views, and suffers from the same difficulties. Propagating changes from an extracted representation back into the original representation is difficult. While some of the solutions proposed for the view update problem may be of use here, we believe that the designer will need to help propagate the changes.

A possible approach is based on having the same set of operations for each representation. While the implementations may vary for each, the operation's abstract function is the same. A log of the operations applied to one representation could then be executed against its equivalent representation, thus keeping them consistent. When some information is lost during the extraction process, the designer can aid in replacing the lost information, or default values can be applied.

7.2 Summary

In this book, we have described the information management needs of computer-aided design systems. Design is an inherently complex acitivity, made even more complex because of the vast quantity of information that must be handled in order to complete the design. We have reviewed the process of design and the sources of complexity in design. The data management requirements for design systems were reviewed, and we showed that existing database systems are inadequate to the task of providing the information management needs of design systems. Part of the complexity is due to the large number of representations that need to be supported, and we illustrated this with examples from the domain of VLSI. A paradigm for how to access a design database and keep it consistent across these representations was presented, as was an overall system architecture.

By now it should be clear to the reader that *design databases* are very *different from commercial databases*. Approaches which have proven to be successful in building applications systems on top of existing database systems will not necessarily work in the design environment. The performance demands, style of interaction, and tolerance for system failure are very different.

A purpose of this book has been to show that the real issue is *how to provide effective design information management*. This involves developing solutions to the problems of controlled sharing of design data, version control, and design consistency maintenance. It is much more complicated than merely storing design data in a database.

We have emphasized the concept of a *design transaction* as a paradigm for how to interact with a design database. The database transaction concept is certainly one of the database research community's most important contributions to computer science to date. Extending it to the design environment has not been straightforward. This is in large part due to the different nature of interaction in the design environment as opposed to conventional transaction processing. While we have described extensions – such as nested subtransactions – to better handle the way in which data is shared in the design environment, it must be admitted that the resulting mechanism is not as elegant as the original. It may well be the case that the transaction concept has been generalized to the point that it is no longer useful, and that a new mechanism must be found to support design interactions. More work is needed in this area.

Chapter 7. Conclusions

The most significant contribution database systems can make in the design environment is *to provide a reliable/durable repository for design data*. Most conventional file systems do not provide the right mechanisms for implementing reliable update of files, which is critical when the data stored is as valuable as design data. We have argued that the recovery mechanisms must be extended beyond the shared repository to the designer' workspaces (continuous recovery) and to structuring of the design data by applications in memory (in-memory recovery).

Finally, from the standpoint of design description, we believe that it is very important to *separate the structural information about the design from the representational information*. Information about how a system is composed from its subsystems, what its interface is, how it has evolved over time, etc., are structural concerns, and fall under the management of the esign management system. On the other hand, the details of the layout or schematic representations fall into the domain of the tools that will create and manipulate this data, and are outside the concern of the design management system.

In conclusion, computer-aided design systems will emerge as one of the most important software systems to be built in the next decade, in the same way that database systems, operating systems, and programming systems were important in the past. Such systems cannot be successful without a consistent data management approach. As a research area within database management, design systems will continue to be a fruitful area for further investigation into how existing techniques should be applied, as well as a driving force for the development of new techniques.

8 Annotated Bibliography

[AYRE83] Ayres, R. F., VLSI – *Silicon Compilation and the Art of Automatic Microchip Design,* Prentice Hall, Englewood Cliffs, N.J, (1983)

Describes a language based approach to VLSI design specification and automatic layout generation

[BAND75] Bandurski, A. E., D. K. Jefferson, "Data Description for Computer-Aided Design", Proc.ACM SIGMOD Conference, (May 1975)

Using a CODASYL database system for a ship design application

[BEET82] Beetem, A., et. al., "Performance of Database Management in VLSI Design", *IEEE TC on DatabaseEngineering Newsletter,* V 5, N 2, (June 1982)

A study of whether commercial database systems can provide adequate performance for VLSI CAD aplications. The answer is a guarded "yes", and depends critically on a proper choice of database structure

[BOBR82] Bobrow, D., M.Stefik, "The LOOPS Manual: A Data and Object Oriented Programming System for InterLisp", Knowledge-Based VLSI Design Group Memo KB-VLSI-81-13, Xerox Palo Alto Research Center, (August 1982)

Describes the use of design libraries based on layers and environments

[BROW83] Brown, H., C. Tong, G. Foyster, "Pallidio: An Exploratory Environment for Circuit Design", *IEEE Computer Magazine,* V 16, N 12, (December 1983)

A VLSI Design Expert System

[CANE83] Canepa, M., E. Weber, H. Talley, "VLSI in FOCUS: Designing a 32-bit CPU Chip", *VLSI Design,* January/February 1983

The Hewlett-Packard 450,000 transistor microprocessor

[CHU83] Chu, K-C, et. al., "VDD – A VLSI Design Database System", Proc. ACM SIGMOD Conference on Engineering Design Applications, San Jose, CA, (May 1983)

A database-oriented VLSI design system developed at AT&T Bell Laboratories

[CIAM76a] Ciampi, P. L., et. al., "Control and Integration of a CAD Database", Proc. 13th ACM/IEEE Design Automation Conference, (June 1976)

[CIAM76b] Ciampi, P. L., J. D. Nash, "Concepts in CAD Database Structures", Proc. 13th ACM/IEEE Design Automation Conference, (June 1976)

[DAMP83] Dampney, C. N.G., "Precedency Control and Other Semantic Issues in a Workbench Database", Proc. ACM SIGMOD Conf. on Engineering Design Applications, San Jose, CA, (June 1983)

[DATE81] Date, C. J., *An Introduction to Database Systems,* Third Edition, Addison-Wesley, Reading, MA, 1981

[DATE82] Date, C. J., *An Introduction to Database Systems,* Vol. II, Addison-Wesley, Reading, MA, 1982

These two volumes are ecxellent general texts on database technology and systems

[EAST80] Eastman, C. M., "System Facilities for CAD Databases, "Proc. 17th ACM/IEEE Design Automation Conference, Minneapolis, MN, (June 1980)

[EAST81a] Eastman, C. M., "Recent Developments in Representation in the Science of Design", Proc. 18th ACM/IEEE Design Automation Conference, (June 1981)

[EAST81b] Eastman, C. M., "Database Facilities for Engineering Design", *Proc. IEEE,* V 69, N 10, (October 1981)

A series of papers by Charles Eastman providing a good overview of the requirements for CAD databases and the state-of-the-art in CAD for architecture

[EDMO83] Edmond, J. C., G. Marechal, "Experience in Building ARCADE, a Computer Aided Design System Based on a Relational DBMS", Proc. ACM SIGMOD Conf. on Engineering Design Applications, San Jose, CA, (June 1983)

Chapter 8. Bibliography

[ESWA76] Eswaren, K. P., J. N. Gray, R. A. Lorie, I. L. Traiger, "The Notions of Consistency and Predicate Locks in a Database System", *Comm. ACM,* V 19, N 11, (November 1976)

[FELD79] Feldman, S. J., "MAKE – A Program for Maintaining Computer Programs", UNIX Time-Sharing System UNIX Programmer's Manual, Seventh Edition, Volume 2A, (January 1979)

> A description of a UNIX utility for maintaining consistent configurations of user programs. If file A depends on file B, then it must have a newer timestamp. If it does not, a script of UNIX commands is invoked to create a new version of file A

[FITC82] Fitch, A., session chair, "Automation in the Creation of the IBM 3081", Proc. 19th ACM/IEEE Design Automation Conference, Las Vegas, Nv, (June 1982)

> IBM used a simulation-based design approach for the 3081, which is described here. Interesting statistics are given about how many logic bugs were caught during the simulation effort

[FOST75] Foster, J. C., "The Evolution of an Integrated Database", Proc. 12th ACM/IEEE Design Automation Conference, (June 1975)

[GRAY78] Gray, J. N., "Notes on Database Operating Systems", IBM San Jose Research Report # RJ2188(30001), (February 1978)

[GRAY81a] Gray, J. N., et al., "The Recovery manager of the System-R Data Manager", *ACM Computing Surveys,* V 13, N 2, (June 1981)

[GRAY81b] Gray, J., "The Transaction Concept: Virtues and Limitations", Proc. 7th Intl. Conf. on Very Large Databases, Cannes, France, (September 1981)

> An excellent paper on the strengths and weaknesses of the database transaction concept

[GUTT82] Guttman, A., M. Stonebraker, "Using a Relational Database Management System for Computer Aided Design Data", *IEEE TC on Database Engineering Newsletter,* V 5, N 2, (June 1982)

> A performance evaluation of using the INGRES relational database system as a CAD database. The results were not very encouraging for the unaugmented system

[HALL84] Hallmark, G., R. A. Lorie, "Towards VLSI Design Systems Using Relational Databases", Proc. IEEE Spring Computer Conference, San Francisco, CA, (February 1984)

> A description of how a geometry editor ha been interfaced to the extended System-R relational database system

[HASK82a] Haskin, R., R. Lorie, "On Extending the Functions of a Relational Database System", Proc. ACM SIGMOD Conference, Orlando, FL, (June 1982)

> Describes the enhancements made to System-R to make it a better basis for CAD applications. The enhancements include: long duration transactions, complex objects, and long data fields

[HASK82b] Haskin, R., R. Lorie, "Using a Relational Database System for Circuit Design", *IEEE TC on Database Engineering Newsletter,* V 5, N 2, (June 1982)

> A further description of the enhancements made to System-R

[HAYE83] Hayes-Roth, F., D. A. Waterman, D. B. Lenat, *Building Expert Systems,* Addison-Wesley, Reading, MA, (1983)

> What expert systems are and how to build them

[HAYN81] Haynie, M. N., "The Relational/Network Hybrid Data Model for Design Automation Databases", Proc. 18th ACM/IEEE Design Automation Conference, (June 1981)

> An attempt to define a good data model for design description by combining the best features of relational (tabular) and network (graph) structured data models

[HOLL84] Hollaar, L., B. Nelson, T. Carter, R.A. Lorie, "The Structure and Operation of a Relational Database System in a Cell-Oriented Integrated Circuit Design System", Proc. 21st ACM/IEEE Design Automation Conference, Alburquerque, NM, (June 1984)

> A description of a cell-based design system that is being interfaced to the System-R relational database system

[HOSK79] Hoskins, E. M., "Descriptive Databases in Some Design/Manufacturing Environments", Proc. 16th ACM/IEEE Design Automation Conference, (June 1979)

[JOHN82] Johnson, H. R., D. L. Bernhardt, "Engineering Data Management Activities within the IPAD Project", *IEEE TC on Database Engineering Newsletter,* V 5, N 2, (June 1982)

> The IPAD project is developing a standard design database system for the aerospace industry. The issues in building such a system, and their progress to date, are reported in this article

[JOHN83] Johnson, H. R., et. al., "A DBMS Facility for Handling Structured Engineering Entities", Proc. ACM SIGMOD Conference on Engineering Design, San Jose, CA, (June 1983)

> A further elaboration of the structure of the IPAD design system

[KATZ82] Katz, R. H., "A Database Approach for Managing VLSI Design Data", Proc. 19th ACM/IEEE Design Automation Conference, Las Vegas, NV, (June 1982)

> A proposed approach for how to use conventional database technology for VLSI CAD

[KATZ83a] Katz, R. H., "Managing the Chip Design Database", *IEEE Computer,* V 16, N 12, (December 1983)
   The description of the design data model of Chap. 4 is an elaboration of the ideas in this paper
[KATZ83b] Katz, R. H., S. Weiss, "Chip Assemblers: Concepts and Capabilities", Proc. 20th ACM/IEEE Design Automation Conference, Miami, FL, (June 1983)
   A description of an information management oriented design tool
[KATZ83c] Katz, R. H., "DAVID: Design Aids for VLSI using Integrated Databases", *IEEE TC on DatabaseEngineering Newsletter,* V 5, N 2, (June 1983)
[KATZ83d] Katz, R. H., "Transaction Management in the Design Environment", Proc. Special Workshop on New Applications of Databases, Cambridge, England, (September 1983), also in *New Applications of Databases,* E. Gelenbe, G. Gardarin, eds., Academic Press, London, 1984
   Chapter 5 is based on this paper
[KATZ84] Katz, R. H., W. Scacchi, P.Subrahmanyam, "Development Environments for VLSI and Software", *Journal of Systems and Software,* to appear
   A comparative evaluation of the state-of-the-art in design environments for VLSI and for software
[KAWA78] Kawano, et. al., "The Design of a Database Organization for an Electronic Equipment Design Automation System", Proc. 15th ACM/IEEE Design Automation Conference, (June 1978)
[KELL82] Keller, K. H., A. R. Newton, S. Ellis, "A Symbolic Design System for Integrated Circuits", Proc. 19th ACM/IEEE Design Automation Conference, Las Vegas, NV, (June 1982)
   A sticks-based design system that has been built on top of a file-oriented design management component
[KIDD81] Kidder, T., *The Soul of a New Machine,* Little, Brown, and Co., Boston, 1981
   The real man's approach to computer engineering
[KIM83] Kim,W., et. al., "Nested Transactions for Engineering Design Databases", IBM Tech Report
   A description of a nested transaction mechanism similar to that presented in Chap. 5
[KORE75] Korenjak, A. J., A. H. Tiger, "An Integrated CAD Database System", Proc. 12th ACM/IEEE Design Automation Conference, (June 1975)
[KUTA83] Kutay, A. R., C. Eastman, "Transaction Management in Engineering Databases", Proc. ACM SIGMOD Conference on Engineering Design Applications, San Jose, CA, (May 1983)
   A view of design transactions as data flowing among design applications. Examples are drawn from the architectural design domain
[LORI81] Lorie, R. A., "A Project on Design Systems", *IEEE TC on Dtabase Engineering Newsletter,* V 4, N 1, (September 1981)
   An early discussion of the project to extend System-R for design applications
[LOSL75] Losleben, P., "Data Structures, Data Base, and File Maintenance", in *Digital System Design Automation: Languages, Simulation & Data Base,* M. A. Breuer, ed., Computer Science Press, Rockville, MD, 1975
[LOSL80] Losleben, P., "ComputerAided Design for VLSI", in *Very Large Scale Integration VLSI: Fundamentals and Applications,* D.F.Barbe, ed., Springer Series in Electrophysics 5, Springer Verlag, Berlin, 1980
   Two excellent surveys of design tools, with an unusually strong emphasis on the data management issues
[MCLE83] McLeod, D. K., et. al., "An pproach to Information Management in CAD/VLSI Applications", Proc. ACM SIGMOD Conference on Engineering Design Applications, San Jose, CA, (May 1983)
   A semantic data model approach for describing design data structures
[MELK83] Melkanoff, M. A., Q. Chen, "An Experimental Database Which Combines Static and Dynamic Capabilities", Proc. ACM SIGMOD Conf. on Engineering Design Applications, San Jose, CA, (June 1983)
[MITS80] Mitsuhashi, T., et. al., "An Integrated Mask Artwork and Analysis System", Proc. 17th ACM/IEEE Design Automation Conference, (June 1980)
[MUDG80] Mudge, J. C., et. al., "A VLSI Chip Assembler", in *Design Methodologies for Very Large Scale Integrated Circuits,* Nato Advanced Summer Institut, Belgium, 1980
[MUDG81] Mudge, J. C., "VLSI Chip Design at the Crossroads", in VLSI 81: *Very Large Scale Integration,* J. P. Gray, ed., Academic Press, London, 1981
   Two early discussions of the chip assembler ideas
[NEUM82] Neumann, T., C. Hornung, "Consistency and Transactions in CAD databases", Proc. 8th Intl. Conf. on Very Large Databases, Mexico City, Mexico, (September 1982)

Chapter 8. Bibliography

[NEUM83] Neumann, T., "On Representing the Design Information in a Common Database", Proc. ACM SIGMOD Conference on Engineering Design Applications, San Jose, CA, (May 1983)

 Two papers describing design transactions from a consistency maintenance viewpoint

[NEWT81] Newton, A. R., "Computer-Aided Design of VLSI Circuits", *Proceeding IEEE,* V 69, N 10, (October 1981)

 A very thorough survey of design tools for VLSI design

[NIEN79] Nieng, K-Y, D. A. Beckly, "Component Library for an Integrated Design Automation System", Proc. 16th ACM/IEEE Design Automation Conference, (June 1979)

[NOON82] Noon, W.A., et. al., "A Design System Approach to Data Intergrity", Proc. 19th ACM/IEEE Design Automation Conference, Las Vegas, NV, (June 1982)

 A description of a design audit trail mechanism used at IBM

[POWE83] Powell, M. L., M. A. Linton, "Database Support for Programming Environments", Proc. ACM SIGMOD Conf. on Engineering Design Applications, San Jose, CA, (June 1983)

 Describes the use of a database system (INGRES) in a program development environment. The database not only contains the fine structure of the program (as a parse tree), but also is used to maintain execution traces

[ROBE81] Roberts, et. al., "A Vertically Organized Computer-Aided Design Data Base", Proc. 18th ACM/IEEE Design Automation Conference, (June 1981)

[ROCH75] Rochkind, M. J., "The Source Code Control System", *IEEE Transaction on Software Engineering,* V SE-1, N 4, (December 1975)

 A version control mechanism for files. Widely in use on UNIX systems

[ROSE80] Rosenberg, L. M., "The Evolution of Design Automation to Meet the Challenges of VLSI", Proc. 17th ACM/IEEE Design Automation Conference, (June 1980)

[SEQU83] Sequin, C. H., "Managing VLSI Complexity: An Outlook", *Proc. IEEE,* V 71, N 1, (January 1983)

 A good discussion on why VLSI design is so difficult, and what we can do to keep the design task to a manageable level

[SIDL80] Sidle, T. W., "Weakness of Commercial Database Management Systems in Engineering Design", Proc. 17th ACM/IEEE Design Automation Conference, (June 1980)

 What is wrong with commercial database systems when they are used in the CAD environment

[SOUT83] Southard, J. R., "Mac Pitts: An Approach to Silicon Compilation", *IEEE Computer Magazine,* V 16, N 12, (December 1983)

 A successful silicon compiler system

[STON83] Stonebraker, M.R., et. al., "Application of Abstract Data Types and Abstract Indices to CAD Databases", Proc. ACM SIGMOD Conference on Engineering Design Applications, San Jose, CA, (June 1983)

 A description of how the abstract data type mechanism within INGRES can be used to assist CAD applications

[SUCH79] Sucher, D. J., D. F. Wann, "A Design Aids Database for Physical Components", Proc. 16th ACM/IEEE Design Automation Conference, (June 1979)

[THOM78] Thompson, K., "UNIX Implementation", *The Bell System Technical Journal,* V 57, N 6, Part 2, (July-August 1978)

[THOM83] Thomas, D. E., et. al., "Automatic Data Path Synthesis", *IEEE Computer Magazine,* V 16, N 12, (December 1983)

 A description of the current state of the CMU design automation system, with particular emphasis on the automatic generation of computer data paths

[VALL75] Valle, G., "Relational Data Handling Techniques in IC Mask Layout Procedures", Proc. 12th ACM/IEEE Design Automation Conference, (June 1975)

[WASS81] Wassermann, A. I., *Softwre Development Environments,* IEEE Computer Society Press, Los Alamitos, CA, (1981)

 A good survey of the state-of-the-art in software development environments

[WIED82] Wiederhold, G., et. al., "A Database Approach to Communication in VLSI Design", *IEEE Trans. on Computer-Aided Design,* V CAD-1, N 2, (April 1982)

 How to use commercial database systems for VLSI design, and still obtain acceptable performance

[WILM79] Wilmore, J. A., "The Design of an Efficient Data Base to Support an Interactive LSI Layout System", Proc. 16th ACM/IEEE Design Automation Conference, (June 1979)

[WONG79] Wong, S., W. Bristo, "A Computer Aided Design Database", Proc. 16th ACM/IEEE Design Automation Conference, (June 1979)

[WORK74] Works, K., et. al., "Engineering Data Management Systems (EDMS) for Computer-Aided Design of Digital Systems", Proc. 11th ACM/IEEE Design Automation Conference, (June 1974)

[ZINT81] Zintl, G., "A CODASYL CAD Database System", Proc. 18th ACM/IEEE Design Automation Conference, (June 1981)